PETER OF LIMOGES

The Moral Treatise on the Eye

The Moral Treatise on the Eye was a medieval bestseller. Composed by Peter of Limoges at the university in Paris between 1274/75 and 1289, the work was transmitted throughout Europe and is still extant in 219 manuscripts. Paris was the crossroads of vibrant intellectual exchange in Peter's time, and he played an important role in bringing together in his *Treatise* advances in both homiletics and science, adapting for use in the pulpit the new science of Perspectivist optics developed by Alhacen and adopted by Roger Bacon and John Pecham. Peter's *Treatise* helped legitimate the cultural utility of natural science by making it part of a theological process of allegorizing the revealed world. It also supplemented the tasks of homiletics with the duty of educating the Christian senses, making the study of vision part of what can be called a "voluntarist optics," one of the necessary steps in the formation of the will. Yet, in spite of its importance to homiletic discourse, the early development of science in the West, and the conceptualization of perspective in art, Peter's text has long been available only in Latin.

This volume makes *The Moral Treatise on the Eye* available for the first time in English. In fifteen chapters, the text moves from what can be established about the eye and vision on scientific grounds to the moral or spiritual interpretation of optics. The first section lays the foundation for the rest of the treatise by detailing the parts of the eye, the types of physical vision, the importance of intention for the completion of sight in the common nerve, the explanation of optical illusions, and the moral interpretation of twelve natural properties of eyes. In the second part, Peter develops sets of moral instruction using cultural conceptions surrounding the eyes. The final section is devoted to spiritual sight.

The volume also contains an introduction to Peter of Limoges, his intellectual interests, the sources of the *Treatise*, and its place in the theological and scientific developments of the late thirteenth century; a detailed index of the sources and analogues; and a bibliography.

MEDIAEVAL SOURCES IN TRANSLATION

Series Editor

MEDIAEVAL SOURCES IN TRANSLATION 51

PETER OF LIMOGES

The Moral Treatise on the Eye

Translated with an introduction by

RICHARD NEWHAUSER

PONTIFICAL INSTITUTE OF MEDIAEVAL STUDIES

Library and Archives Canada Cataloguing in Publication

Peter, of Limoges, d. 1306
The moral treatise on the eye / Peter of Limoges ; translated with an introduction by Richard Newhauser.

(Mediaeval sources in translation, 0316-0874 ; 51)
Translation of: Tractatus moralis de oculo.
Includes bibliographical references and indexes.
ISBN 978-0-88844-301-4

1. Eye – Early works to 1800. 2. Vision – Early works to 1800. 3. Optics – Early works to 1800. 4. Vision – Moral and ethical aspects – early works to 1800. 5. Catholic Church – Clergy – Training of. I. Newhauser, Richard, 1947- II. Pontifical Institute of Mediaeval Studies III. Title. IV. Series: Mediaeval sources in translation ; 51

QP475.P4813 2012 612.8'4 C2011-907914-3

Pontifical Institute of Mediaeval Studies
59 Queen's Park Crescent East
Toronto, Ontario, Canada M5S 2C4

www.pims.ca

MANUFACTURED IN CANADA

Contents

Acknowledgements

A number of people have helped in the completion of this translation and I wish to thank them for their support. Students at Arizona State University, Tempe, assisted in reviewing the Introduction, creating part of the index, or collating changes between *P7P8* and *Pr1*: I am grateful for the work of Nathaniel Bump, Constance Trent, Janna Arenas, Kasandra Castle, and Alysia Santellan. For all things concerning the science of optics, I have been fortunate to have the advice of A. Mark Smith, University of Missouri; for advice on Peter's Hebrew etymologies, I am grateful to Francoise Mirguet, Arizona State University, Tempe. For her care in reading, commenting on, and suggesting changes to drafts of the translation, I acknowledge the diligence of Bridget K. Balint, University of Indiana. The indefatigable efforts of Winston E. Black, University of Tennessee, were also essential to improving the translation. Suzanne Conklin Akbari, University of Toronto, made a number of valuable suggestions that have strengthened the book, and I have also been fortunate to have enjoyed the enthusiastic support of Herbert L. Kessler, Johns Hopkins University. Without the encouragement of the series editor, Mary Carruthers, All Souls College, Oxford, this book would never have been begun and certainly not completed. At PIMS Press, I have had the privilege to work with Jean Hoff, the consummate editor, whose tireless work has not only enhanced every aspect of this translation, but also sets an example of precision for all editors to emulate. My wife, Andrea Németh-Newhauser, has advised me on countless issues of translation technique, an area in which her expertise has improved the rendering of Peter of Limoges' treatise. Finally, my interest in *The Moral Treatise on the Eye* was supported earliest, and most importantly, by Fr. Louis-Jacques Bataillon, OP, whose generosity in sharing with me his knowledge of Peter of Limoges and whose encouragement to complete the translation is gratefully acknowledged here. He did not live to see it published, but whatever in it is worth a reader's time was inspired by him.

Abbreviations

CCCM	Corpus Christianorum. Continuatio Medieualis. Turnhout, Belgium: Brepols.
CCSL	Corpus Christianorum. Series Latina. Turnhout, Belgium: Brepols.
CFS	Cistercian Fathers Series. Kalamazoo, MI.
CSEL	Corpus Scriptorum Ecclesiasticorum Latinorum. Vienna.
FOTC	The Fathers of the Church. Washington, DC.
LCL	Loeb Classical Library. Cambridge, MA, and London.
MGH	Monumenta Germaniae Historica. Hannover etc. Available at http://www.dmgh.de/ (accessed August 1, 2011).
MST	Mediaeval Sources in Translation. Toronto.
NPNF	The Nicene and Post-Nicene Fathers. 1: First Series: 14 vols. 2: Second Series: 14 vols. Buffalo, NY: Christian Literature Publishing Co., 1886–1900. Rev. ed. for Internet Kevin Knight. Available at http://www.newadvent.org/fathers/ (accessed August 1, 2011).
P7	Paris, Bibliothèque nationale de Paris MS. lat. 16395.
P8	Paris, Bibliothèque nationale de Paris MS. lat. 16396.
PG	Patrologiae Cursus Completus. Series Graeca. Ed. Jacques-Paul Migne. 161 vols. Paris: Jacques-Paul Migne, 1857–1866.
PL	Patrologiae Cursus Completus. Series Latina. Ed. Jacques-Paul Migne. 221 vols. Paris: Garnier, 1844–1898.
Pr1	Peter of Limoges. *Tractatus moralis de oculo*. In *Johannis Pithsani archiepiscopi Canthuariensis ordinis fratrum minorum liber de oculo morali foeliciter incipit*. Augsburg: A. Sorg, 1475 [?].
WSA	The Works of Saint Augustine: A Translation for the 21st Century. Hyde Park, NY.
*	An asterisk in the translation indicates an emendation of the Latin text or a significant variation between the text of *P7P8* and *Pr1*; see below, "Emended Readings and Significant Variations between *Pr1* and *P7P8*."

Biblical books referred to in the footnotes are listed below. Abbreviations for other primary sources cited in the footnotes are found below in the "Index of Sources and Analogues."

Gen	Genesis
Exod	Exodus
Lev	Leviticus
Num	Numbers
Deut	Deuteronomy
Judg	Judges
1Sam	1 Samuel
2Sam	2 Samuel
1Kgs	1 Kings
2Kgs	2 Kings
2Chr	2 Chronicles
Tob	Tobit
Jth	Judith
Esth	Esther
Job	Job
Ps	Psalms
Prov	Proverbs
Qoh	Qohelet [Ecclesiastes]
Cant	Canticles [Song of Songs]
Wis	Wisdom
Sir	Sirach [Ecclesiasticus]
Isa	Isaiah
Jer	Jeremiah
Lam	Lamentations
Bar	Baruch
Ezek	Ezekiel
Dan	Daniel
Hos	Hosea
Joel	Joel
Amos	Amos
Mic	Micah
Nah	Nahum
Zech	Zechariah
Mal	Malachi
1Mac	1 Maccabees
2Mac	2 Maccabees
Matt	Matthew
Mark	Mark

Luke	Luke
John	John
Acts	Acts of the Apostles
Rom	Epistle to the Romans
1Cor	First Epistle to the Corinthians
2Cor	Second Epistle to the Corinthians
Gal	Epistle to the Galatians
Eph	Epistle to the Ephesians
Phil	Epistle to the Philippians
Col	Epistle to the Colossians
2Thess	Second Epistle to the Thessalonians
1Tim	First Epistle to Timothy
2Tim	Second Epistle to Timothy
Tit	Epistle to Titus
Heb	Epistle to Hebrews
Jas	Epistle of James
1Pet	First Epistle of Peter
2Pet	Second Epistle of Peter
1John	First Epistle of John
Rev	Revelation

Introduction

As today, in the Middle Ages the name "Peter of Limoges" was hardly a household item. Outside the immediate environment of Paris, his name was often unknown, even though the work translated here was among the more important influences on the composition of sermons and other homiletic texts in the later Middle Ages and played a major role in the transmission of the new science of optics to a popular audience in that period. In England, for example, where *The Moral Treatise on the Eye* was copied often and quoted in numerous sermons, there is only ambiguous evidence that Peter's name was ever known.[1] Nevertheless, Peter was an active contributor to the intellectual life of the university in Paris, and the importance of his treatise is confirmed by the large number of its still extant copies, the many references to the work in sermons, and the frequent borrowings from it found in other texts throughout the late Middle Ages. It is to be hoped that this translation will contribute to the growing recognition, among modern readers, of Peter's achievement and the place of his treatise in intellectual history.

Born in the first half of the thirteenth century in Donzenac in the vicinity of Brives (diocese of Limoges), Peter was active in Paris at a time of sweeping developments in theology, philosophy, and science, espe-

1 In a single early codex, Durham, University Library MS. Mickleton & Spearman 89 (s. xiii ex), one finds an attribution to <... *d>e lymochia* on fol. 1r (the folio is torn), but that notation is in a fifteenth-century hand. The manuscript is from England and was in the library of the Benedictine cathedral priory of St. Cuthbert in Durham in the late fifteenth–early sixteenth century (it is signed by Henry Thew, monk in Durham from ca. 1483 into the 1530s, on fol. iv[v]). Two other English copies of the text attribute the work to *magister de Limochia* (or *Lunochia*) in the fifteenth century: Oxford, Bodleian Library MS. Bodl. 315 (*SC* 2712), and Oxford, Bodleian Library MS. Bodl. 742 (*SC* 2756). But it is unclear what location Lymochia, etc., was thought to designate. One manuscript now in England carries an attribution to Peter of Limoges (*magister Petrus de Cyperia Lemouicensis dyocesis*), but it is a fifteenth-century codex from Germany: Oxford, Bodleian Library MS. Hamilton 21 (*SC* 24451).

cially optics. He died on 2 November 1306.[2] According to the necrologium of the Sorbonne, Peter achieved the baccalaureate in theology, was a canon in Évreux, and had a reputation as a well-known astronomer.[3] He may have completed the *magister artium* by 1262,[4] but his university dossier shows that he must have completed his theology studies with the *magister* degree as well.[5] His university career might also have included functioning as dean of the medical school in 1267 and again in 1270.[6] Peter is further noted for twice having refused prebends in Paris and twice having declined election to the office of bishop. His reasons for rejecting these honors are not hinted at in the necrologium, but perhaps his disaffection with multiple benefices motivated him to some degree since this was a contemporary ecclesiastical abuse he criticizes frequently in his treatise. But he was also affluent enough on his own not to require the extra income these honors would have resulted in: as the necrologium mentions further, Peter bequeathed to the Sorbonne his entire library of more than 120 volumes, which in the early fourteenth century was a significantly large and expensive collection for an individual and argues for the extensive nature of his personal wealth.[7] His

2 Bernard Hauréau, "Sermonnaires," in *Histoire littéraire de la France* (Paris, 1873), vol. 26, 387–468, here 460–63; see also Hieronymus Spettmann, "Das Schriftchen 'De oculo morali' und sein Verfasser," *Archivum Franciscanum Historicum* 16 (1923): 309–22, here 317–22.

3 For Peter's activity as an astronomer, see Lynn Thorndike, "Peter of Limoges on the Comet of 1299," *Isis* 36 (1945): 3–6.

4 Palémon Glorieux, *Aux origines de la Sorbonne, I: Robert de Sorbon: l'homme, le collège, les documents*, Études de philosophie médiévale, 53 (Paris, 1966), 323; idem, *Répertoire des maîtres en théologie de Paris au XIIIe siècle*, Études de philosophie médiévale, 17 (Paris, 1933), vol. 1, 366, no. 178.

5 Paris, Bibliothèque nationale de Paris MS. lat. 16390, fols. 9r–14v. See also Nicole Bériou, "La prédication au béguinage de Paris pendant l'année liturgique 1272–1273," *Recherches augustiniennes* 13 (1978): 105–229, here 108 and n13.

6 *Chartularium Universitatis Parisiensis*, ed. Heinrich Denifle and Emile Chatelain (Paris, 1889), vol. 1, 468, 488. It is not clear whether the Peter of Limoges referred to in these entries is identical with the author of *The Moral Treatise on the Eye*.

7 The entry from the necrologium is printed in Léopold Delisle, *Le Cabinet des Manuscrits de la Bibliothèque impériale*, Histoire Générale de Paris, 6 (Paris, 1874; reprint New York, 1973), vol. 2, 168: "Obiit magister Petrus de Lemovicis, quondam socius domus, canonicus Ebroicensis, qui refutavit duos episcopatus et bis prebendam Parisiensem, bacalarius in theologia, magnus astronomus, qui legavit domui plus quam vi^{xx} volumina, exceptis quibusdam caternis, magne reputationis."

bequest is in fact the "second largest known gift to the medieval library" of the Sorbonne.[8]

His career as a member of the faculty of the Sorbonne is reflected in a number of passages in the treatise on the eye. Paris figures here as the setting for *exempla* concerned with university life: as a center of learning for anyone seeking preferment in ecclesiastical office (chapt. 6.2, in a narrative adopted from John of Salisbury's *Policraticus*), or as the site of famous schools of theology (chapt. 6.11).[9] The importance of these passages lies in the fact that the connection of Paris with the university is, in fact, the only context in which the city appears in the treatise. Peter's critical view of student life at the Sorbonne is apparent on a number of occasions. All of chapter 11 is concerned with instruction for scholars, and among these details Peter narrates two vignettes in the first person. The first concerns a well-known type of student:

> I had a companion who was able to stay up the entire night talking about idle and frivolous matters, but who nevertheless, as he frequently confessed to me, so detested his studies that if he were perhaps unable to sleep at some point, he would open a book and after reading four or five lines he would fall asleep immediately.[10]

The second, found later in the same section of the text, is equally critical of the study habits of students:

> A certain layman who was a close acquaintance of mine once asked me in which faculty more students were enrolled at Paris, and I

8 Richard H. Rouse, "The Early Library of the Sorbonne," *Scriptorium* 21 (1967): 42–71, 227–51; reprint in Mary A. Rouse and Richard H. Rouse, *Authentic Witnesses. Approaches to Medieval Texts and Manuscripts*, Publications in Medieval Studies, 17 [27] (Notre Dame, 1991), 341–408, here 376.

9 Though not always numbered sequentially in the text, Peter divides the longer chapters into sections (e.g., Chapter 6 presents "thirteen extraordinary phenomena"). "Chapt. 6.11" refers to the eleventh of these subdivisions.

10 I quote the text from the earliest printed edition, which is the basis of this translation: *Johannis Pithsani archiepiscopi Canthuariensis ordinis fratrum minorum liber de oculo morali foeliciter incipit* (Augsburg, 1475[?]), 11.5: "Habui vnum socium qui poterat tota nocte de ociosis et friuolis fabulando vigilare, qui tamen, vt mihi confessus est sepe, studium habuit sic exosum vt, si contingeret quod alias dormire non posset, aperto libro et perlectis in eo quatuor aut quinque lineis protinus obdormiuit." This edition, unpaginated, is referred to from here on as *Pr1*.

> responded that this was in the Arts or in Logic. And he said, "On the contrary, rather in Music (*musica*), since more of them grumble about things (*mussitant*) than study."[11]

Whether the passages are borrowed as conventional sentiments or not, these narratives are presented as self-representing details from the author's life which place him in the milieu of the Parisian university.

From Peter's private library more than 70 manuscripts are still extant today, a number of them entirely or partially in his own hand.[12] The topics they represent bear witness to an intellectual of wide-ranging interests: works on astrology, astronomy, natural sciences, medicine, as well as theology and collections of *exempla*. The breadth of these interests is also reflected in the wide range of works and writers cited in *The Moral Treatise on the Eye*, often (though by no means always) by name, as only a very few examples will document: besides such well-known authors as Augustine of Hippo, Bernard of Clairvaux, or Seneca, as authorities on rhetoric Peter quotes Geoffrey of Vinsauf and Quintilian; on grammar, Donatus; on medicine, Urso Salernitanus and the *Flos medicinae Scholae Salerni*; on architecture, Vitruvius; and on optics, Alhacen and Euclid. In the field of hagiography, Peter quotes lives of Edmund of Canterbury, John the Almoner, and John the Anchorite; in Scholastic theology, a number of commentaries on Aristotle; in historiography, Helinand of Froidmont; in moral-satirical literature, Johannes de Hauvilla and John of Salisbury. As the reader will understand from the bibliography to this volume, the list of authorities cited by Peter of Limoges could continue at much greater length.

And yet, with all Peter's attention to naming his sources, some authors' names do not occur anywhere in the treatise. For example, he never once acknowledges Roger Bacon as an influence on his text, though as will be seen Bacon's impact on the *Treatise* was greater than any other author's. Peter's suppression of Bacon's name may be due to

11 *Moral Treatise*, 11.5 (*Pr1*): "Quesiuit semel a me quidam laycus mihi familiaris et notus in qua facultate essent plures scolares studentes Parisius. Et respondi quod in artibus vel in loyca. Et ipse: 'Imo, et in musica, quia plures mussitant quam student.'"

12 The known MSS. still extant from Peter's library were first listed by Delisle, *Le Cabinet*, 168. Fr. Louis-Jacques Bataillon placed his updated list of these MSS. at my disposal. On the MSS. written in Peter's own hand, see esp. Madeleine Mabille, "Pierre de Limoges, copiste de manuscrits," *Scriptorium* 24 (1970): 45–47 and illus. 10–13, and eadem, "Pierre de Limoges et ses méthodes de travail," in *Hommages à André Boutemy*, ed. Guy Cambier, Collection Latomus, 145 (Brussels, 1976), 244–51.

Bacon's low reputation among his fellow Franciscans, because of his bitter struggles with his superiors, in the period in which Peter was composing the *Treatise*. At a point when Bacon was very much out of favor, it would hardly have been circumspect to tout his influence on the *Treatise* among those whom Peter must have hoped would use his work in the preparation of their sermons.[13] But Peter also mined texts by other authors who remained unnamed perhaps because as near contemporaries they were not to be treated as *auctoritates*. One important example of this phenomenon is Vincent of Beauvais, whose vast encyclopedia of doctrinal, historical, and natural information, the *Speculum maius*,[14] was Peter's main source for narratives taken from the early-thirteenth-century *Chronicon* by the Cistercian Helinand of Froidmont.[15] Peter cites Helinand by name, but it is most likely that he knew Helinand's work only through the *Speculum maius*. And a number of the other named sources in *The Moral Treatise on the Eye* were probably taken from Vincent's work as well, including Seneca (though Peter knew other sections of Seneca's work directly) and the Emperor Constantine.[16] One can discover Peter's reading of Vincent of Beauvais not only in the close verbal relationships of the two texts. In one case, for example, Peter narrates an episode from the life of John the Anchorite which he has taken from Vincent's *Specu-*

13 See Richard Newhauser, "*Inter scientiam et populum*: Roger Bacon, Peter of Limoges, and the 'Tractatus moralis de oculo,'" in *Nach der Verurteilung von 1277. Philosophie und Theologie an der Universität von Paris im letzten Viertel des 13. Jahrhunderts. Studien und Texte / After the Condemnation of 1277. Philosophy and Theology at the University of Paris in the Last Quarter of the Thirteenth Century. Studies and Texts*, ed. Jan A. Aertsen, Kent Emery, Jr., and Andreas Speer, Miscellanea Mediaevalia, 28 (Berlin and New York, 2001), 682–703, here 693–95.

14 On Vincent of Beauvais and the *Speculum maius*, see Monique Paulmier-Foucart, with Marie-Christine Duchenne, *Vincent de Beauvais et le grand miroir du monde*, Témoins de notre histoire, 10 (Turnhout, 2004); *Lector et compilator. Vincent de Beauvais, frère prêcheur: un intellectuel et son milieu au XIIIe siècle*, ed. Serge Lusignan and Monique Paulmier-Foucart, with Marie-Christine Duchenne (Grâne, France, 1997). See also the web site devoted to Vincent: http://www.vincentiusbelvacensis.eu/ (accessed October 16, 2011).

15 For examples, see below, chapt. 2, pp. 9–10; chapt. 8.4, pp. 88–89; chapt. 9, p. 109. For Helinand, see the recent work of Beverly Mayne Kienzle, for example: "Cistercian Views of the City in the Sermons of Hélinand of Froidmont," in *Medieval Sermons and Society: Cloister, City, University. Proceedings of the International Symposia at Kalamazoo and New York*, ed. Jaqueline Hamesse et al., Textes et études du moyen âge, 9 (Louvain-la-Neuve, 1998), 165–82. See also the web site devoted to Helinand: http://www.vincentiusbelvacensis.eu/home_hf.htm (accessed May 15, 2011).

16 See below, chapt. 6.12, p. 40 (Seneca); chapt. 12.1, pp. 146–47 (Constantine).

lum historiale. Vincent does not name the source of the misogynistic *exemplum* in which John reprimands a woman who has come to see him, but the previous chapter in the *Speculum historiale* ends with material cited by name from Sigebert of Gembloux, and it is probably for this reason that Peter attributes the *exemplum* on John the Anchorite to Sigebert.[17] Peter's treatment of his sources is, thus, more complex than what might appear at first sight.

His private library also contained a number of books by authors associated with the Franciscan Order: a selection of the works of Joachim of Fiore, for example, works by Petrus Johannis Olivi, the *Perspectiva* by Roger Bacon,[18] and treatises by Ramon Llull, whom Peter introduced into the milieu of the Parisian university.[19] Peter also collected works by other original and sometimes unconventional contemporaries, such as the member of the arts faculty Boethius of Dacia, whose work on the eternity of the world was controversial; the Augustinian Giles of Rome; and the Dominican Nicholas of Gorran.[20] Above all, however, Peter collected, and produced, homiletic literature. He copied sermons during their delivery in Paris during the period 1260–1264 and again in 1272–1273, and he himself preached in Paris in 1273 and 1280.[21]

17 See below, chapt. 8.7, p. 103.

18 For manuscripts from Peter's library with treatises by Olivi, see Paris, Bibliothèque nationale de Paris MS. lat. 15559 and MS. lat. 15588; for Bacon's *Perspectiva*, see Paris, Bibliothèque nationale de Paris MS. lat. 7434. Peter's interest in the thought of Joachim of Fiore was examined by Nicole Bériou: "Pierre de Limoges et la fin du temps," *Mélanges de l'École française de Rome* (Série Moyen Age et Temps Modernes) 98,1 (1986): 65–107. For Peter's interest in Petrus Johannis Olivi, see Marie-Thérèse d'Alverny, "Un adversaire de Saint Thomas: Petrus Iohannis Olivi," in *St. Thomas Aquinas 1274–1974: Commemorative Studies*, ed. Armand A. Maurer et al. (Toronto, 1974), vol. 2, 179–218, here 192–206.

19 Albert Soler, "Ramon Llull and Peter of Limoges," *Traditio* 48 (1993): 93–105. The manuscripts from Peter's library with treatises by Llull are Paris, Bibliothèque nationale de Paris MS. lat. 16112, MS. lat. 16113, MS. lat. 16114, MS. lat. 16119, and MS. fr. 24402.

20 For manuscripts from Peter's library with treatises by Boethius of Dacia, see, for example, Paris, Bibliothèque nationale de Paris MS. lat. 16407; for Giles of Rome, see Paris, Bibliothèque nationale de Paris MS. lat. 15863, MS. lat. 16122, MS. lat. 16157, MS. lat. 16616, and MS. lat. 16681. The manuscripts with works by Giles of Rome have been described by Concetta Luna, *Opera omnia, 1.1/3**: Catalogo dei manoscritti (294–372): Francia (Parigi)*, Testi e studi per il "Corpus Philosophorum Medii Aevi," 7 (Florence, 1988), 199–201, 219–20, 226–27, 271–74.

21 Bériou, "Fin du temps," 70 and n22; Jean Longère, *La prédication médiévale* (Paris, 1983), 154, 159, 163; Richard H. Rouse and Mary A. Rouse, *Preachers, Florilegia, and Sermons: Studies on the Manipulus florum of Thomas of Ireland*, Studies and

Paris was the crossroads of vibrant intellectual exchange on a number of fronts at this time, and Peter of Limoges played an important role in bringing together in his treatise advances in both science and homiletics. As David Lindberg has written, one of the tasks of Western intellectuals in the thirteenth century was "to come to terms with the contents of the newly translated texts [of Greek and Arabic philosophy and natural science] – to master the new knowledge, organize it, assess its significance, discover its ramifications, work out its internal contradictions, and apply it (wherever possible) to existing intellectual concerns."[22] The study of optics, specifically, had maintained a longstanding place in studies at the university in Paris since the 1220s, and it had been closely associated with scholars like Robert Grosseteste, that is to say, with theologians who were also interested in natural philosophy. Nevertheless, the last third of the thirteenth century was a period of significant change in the study of vision, light, and visual perspective because it was in this period that Perspectivist optics had its greatest impact on theology in the West, absorbing the work of the great Muslim savant, Alhacen (Ibn al-Haytham) (965–1040).[23] Roger Bacon, one of the most important scholars of Greek and Arabic texts on optics in the middle decades of the thirteenth century,[24] was influenced by Alhacen, as was Peter of Limoges, both directly and through the work of Bacon and that of John Pecham, the author of an important work on Perspectivist optics and from 1279 the archbishop of Canterbury.

Texts, 47 (Toronto, 1979), 78–79; Bériou, "Prédication," 107–16; Mabille, "Copiste," 46–47; Albert Lecoy de la Marche, *La chaire française au moyen âge* (Paris, 1886), 100.

22 David C. Lindberg, *The Beginnings of Western Science. The European Scientific Tradition in Philosophical, Religious, and Institutional Context, 600 B.C. to A.D. 1450*, 2nd ed. (Chicago and London, 2007), 225.

23 David C. Lindberg, "The Science of Optics," in *Science in the Middle Ages*, ed. David C. Lindberg (Chicago and London, 1978), 338–68, here 345ff., reprint in David C. Lindberg, *Studies in the History of Medieval Optics*, Variorum Reprints, Collected Studies Series CS186 (London, 1983), I; idem, *Theories of Vision from Al-Kindi to Kepler* (Chicago and London, 1976), 58–86; idem "Alhazen's Theory of Vision and Its Reception in the West," *Isis* 58 (1967): 321–41, reprint in *Studies in the History of Medieval Optics*, III.

24 For a recent examination of the cultural importance of Bacon's work, see Stephen G. Nichols, "'The Pupil of Your Eye': Vision, Language, and Poetry in Thirteenth-Century Paris," in *Rethinking the Medieval Senses. Heritage, Fascinations, Frames*, ed. Stephen G. Nichols, Andreas Kablitz, and Alison Calhoun (Baltimore, 2008), 286–307. For Bacon as a scientist, see *Roger Bacon and the Sciences: Commemorative Essays*, ed. Jeremiah Hackett, Studien und Texte zur Geistesgeschichte des Mittelalters, 57 (Leiden and New York, 1997).

Pecham's influence can be noted in a number of sections of Peter's work, and his *Perspectiva communis* also had a verbal impact on *The Moral Treatise on the Eye*, for example in Peter's discussion of refraction among the optical illusions explained and moralized in chapter 6.[25] Peter was familiar with all parts of Roger Bacon's *Opus maius*: in chapter 7, his calculation of the size of the sun relative to that of the earth may have been taken from Bacon's citation of the astronomer Thabit ibn Qurra on this matter in book 4 of the *Opus maius*, and a citation from Seneca in chapter 11 of Peter's work may have been borrowed from book 7 of Bacon's text.[26] But it is especially Bacon's *Perspectiva*, which also became part of the *Opus maius*, that was formative for *The Moral Treatise on the Eye*.[27] Peter not only borrowed the details of optical science from Bacon and the interpretation of the work of earlier writers on optics, as he does, for example, in chapter 11, where he adds to a discussion found in Ptolemy the necessity of a duration of time as a factor allowing for the formation of visual judgment, precisely as Roger Bacon had done.[28] But even more important, in the section of the *Perspectiva* that is devoted to the utility of Perspectivist science for theology, part of Bacon's far-reaching plans for a reformation of all of Christian education, Peter discovered the very foundations for his own work. Book 3 of the *Perspectiva* provided him with the necessary theoretical support to bring together the science of optics and moral theology and to make of them a unified discourse, fit for presentation in the pulpit.[29]

Sight and the functions of the eye were the central issues for those engaged in Perspectivist optics, rather than theories of light and radiation in their own right, and in this way the work of the Perspectivists on vision developed a science of visual cognition that explained by stages the movement from the physical perception of an image by the eye to the cognition of that form in abstract terms.[30] The importance in this process of

25 See below, chapt. 6.4, p. 27. On Peter's debt to John Pecham, see Gudrun Schleusener-Eichholz, "Naturwissenschaft und Allegorese: Der 'Tractatus de oculo morali' des Petrus von Limoges," *Frühmittelalterliche Studien* 12 (1978): 258–309, especially 288–90.

26 See below, chapt. 7.9, p. 65, for the calculation that the sun is 166 times as large as the earth, and chapt. 11.1, p. 117, for a passage borrowed from Seneca's *De ira*.

27 On Peter's debt to Roger Bacon, see Newhauser, "*Inter scientiam et populum*," 682–703.

28 See below, chapt. 11.6, pp. 140–41.

29 See below, Prol., p. 3; chapt. 1, p. 6; and chapt. 4, p. 15; for examples of Peter's borrowings from this section of Bacon's *Perspectiva*.

30 A. Mark Smith, "Getting the Big Picture in Perspectivist Optics," *Isis* 72 (1981): 568–89.

the viewer's evaluation of what was perceived was also central to the Perspectivists,[31] and it was here that works of *moralia* in the Middle Ages such as *The Moral Treatise on the Eye* played an important role. They provided a medium through which perceptions in the natural world could be made comprehensible as demonstrations of what was presented as their spiritual and moral equivalents. Peter's treatise helped legitimate the cultural utility of natural science by making it part of a theological process of allegorizing the revealed world.[32] Furthermore, Peter emphasized far more explicitly than had others before him that the viewer's willful intent in viewing was part of the optical – and moral – process moving from visual sensation to the judgment and evaluation of what has been seen.[33]

An emphasis on the eye in the metaphorical language describing the focus of the faulty will on an object of its desire had not been missing in earlier moral treatises. In the *Summa on the Vices* by the Dominican William Peraldus (completed ca. 1236), for example, one finds a number of passages in which the eye serves as metonymy for the single-minded obsessiveness of the sinner's attachment to sin. In the treatise on avarice in the *Summa*, in a sentiment that Peter of Limoges would echo in the later thirteenth century, Peraldus finds fault with clerics who multiply their benefices. A cleric might try to justify having many benefices by arguing that they amount only to what is sufficient for him, but the real cause of this kind of rationalization for ever more benefices is found to be articulated in Zechariah 5:

> "Their eye in all the earth is" [Zech 5:6] impiety. The eye of one who is lustful is lust, the eye of one who is avaricious is avarice, and so on for the rest of the impieties. It is for this reason that a foolish woman appears beautiful to someone who is lustful and who adores her with excessive love, although in fact she is ugly, and this happens since he

31 Katherine H. Tachau, "Seeing as Action and Passion in the Thirteenth and Fourteenth Centuries," in *The Mind's Eye. Art and Theological Argument in the Middle Ages*, ed. Jeffrey F. Hamburger and Anne-Marie Bouché (Princeton, 2006), 336–59, here 351.

32 On Peter's place in the "hybridization of science and theology," see most recently Herbert L. Kessler, "*Speculum*," *Speculum* 86,1 (2011): 1–41; as well as the earlier work of Schleusener-Eichholz, "Naturwissenschaft." On the eye as metaphor, see Gudrun Schleusener-Eichholz, *Das Auge im Mittelalter*, Münstersche Mittelalter-Schriften, 35 (Munich, 1985).

33 See below chapt. 4, p. 14, and chapt. 11.5, p. 134, for Peter's discussion of intentionality in the visual process.

> looks at her with the eye of lust. For the excessive and lustful love which is within him makes him judge her beauty in this way, but another person who does not have this kind of eye does not see beauty in that woman. Thus, by a false or fantasy-ridden vision someone in whom there is the eye of avarice sees that a person can hold several ecclesiastical benefices requiring service to the church, but those without the eye of avarice do not see this.[34]

Peraldus' text shows a number of points of contact with *The Moral Treatise on the Eye* and may have been used by Peter of Limoges. In any case, what Perspectivist optics added to Peter's discussion was the view of a scientific, even physiological, basis for exactly this kind of moral analysis.

The treatise supplemented the tasks of homiletics with the further duty of educating the Christian senses, making the study of vision not just a physical science, but part of what can be called a "voluntarist optics," one of the necessary steps in the formation of the will.[35] In this light, Peter's frequent appeals to the explanatory power of Perspectivist optics can be seen as both a recognition of the fragility and unreliability of sensory perception, and a statement of belief in the perfectibility of the cognitive process under the tutelage of theological pedagogy. Unaided, seeing might not be believing, but with a solid foundation in Perspectivist optics pastoral theologians could help remedy this situation by educating others in how to see.

Yet in the later thirteenth century there was also a price to be paid by the Parisian university's theology faculty for a too keen interest in natural philosophy. The reaction to what was perceived as an excessive interest in Aristotle and the reasoning of natural philosophy included also the Condemnation of 1277, when Stephen Tempier, the bishop of Paris, forbade the teaching of 219 philosophical and theological propositions at the uni-

34 Paris, Bibliothèque Mazarine MS. 794 (1244), fol. 66rb: "Impietas est oculus eorum in uniuersa terra. Luxus oculus est luxuriosi, auaricia est oculus avari, et sic de ceteris impietatibus. Vnde accidit quod alicui luxurioso uidetur aliqua fatua mulier, quam nimio diligit amore, pulchra, cum in ueritate sit turpis, et hoc est quia oculus luxurie eam intuetur. Nimius enim amor luxuriosus, qui in eo est, facit eum sic iudicare de pulcritudine illius. Alius autem, qui non habet huiusmodi oculum, non uidet illam pulcritudinem in illa muliere. Sic ille, in quo est oculus auaricie, uidet quod aliquis posset habere plura beneficia ecclesiastica seruiencia uel visione falsa uel fantastica. Illi autem, qui non habent oculum auaricie, non uident hoc."

35 Richard Newhauser, "Peter of Limoges, Optics, and the Science of the Senses," in *Pleasure and Danger in Perception: The Five Senses in the Middle Ages and the Renaissance*; special issue of *The Senses & Society* 5,1 (2010): 28–44.

versity there.[36] In the face of such reactions and a growing fideism, theologians had also to take care that their interest in natural science would not be construed as the sin of vain curiosity.[37] As much as Peter depends on Perspectivist optics for the very foundation of his treatise, he is careful to note that secular disciplines by themselves contain "the vanity of multiple error," an image of which he finds in the Tower of Babel:

> It is related in Genesis 11 how communication in a unified language was taken away from those who were presumptuous when the tower of Babel was being erected presumptuously into the heavens, and how a partition of peoples followed a separation of languages. So also while the philosophers raised to the heights the devices of their intellect, prying into heavenly and divine matters with an illicit curiosity and presuming to do so by their own powers, the unity of truth, which is both immutable and unfailing, was in fact removed from them. Enveloped by clouds of ignorance and given over to a reprehensible way of thinking, they were dispersed into sects of various types of errors, and "claiming that they were wise, they were made fools," Romans 1.[38]

36 On the Condemnation, see Kent Emery, Jr., and Andreas Speer, "After the Condemnation of 1277: New Evidence, New Perspectives, and Grounds for New Interpretations," in *Nach der Verurteilung von 1277*, ed. Aertsen, Emery, Jr., and Speer, 3–19.

37 For the accusation of sinful curiosity in and around the European university from the twelfth century on, see Neil Kenny, *The Uses of Curiosity in Early Modern France and Germany* (Oxford, 2004), 25–98; Peter Godman, *The Silent Masters: Latin Literature and Its Censors in the High Middle Ages* (Princeton, 2000); Gillian R. Evans, *Getting it Wrong. The Medieval Epistemology of Error*, Studien und Texte zur Geistesgeschichte des Mittelalters, 63 (Leiden, etc., 1998), chapter four; Olga Weijers, *Terminologie des universités au XIIIe siècle*, Lessico intellettuale europeo, 39 (Rome, 1987); Edward Peters, "*Libertas Inquirendi* and the *Vitium Curiositatis* in Medieval Thought," in *La Notion de liberté au Moyen Âge. Islam, Byzance, Occident: Penn-Paris-Dumbarton Oaks Colloquia, IV* (Paris, 1985), 89–98; Stephen C. Ferruolo, *The Origins of the University: The Schools of Paris and Their Critics, 1100–1215* (Stanford, 1985), 231–77; and see also Richard Newhauser, "Towards a History of Human Curiosity. A Prolegomenon to Its Medieval Phase," *Deutsche Vierteljahrsschrift* 56 (1982): 559–75; reprint in Richard Newhauser, *Sin: Essays on the Moral Tradition in the Western Middle Ages*, Variorum Collected Studies Series, CS869 (Aldershot, Engl., and Burlington, VT: Ashgate Publishing, 2007), essay XIII.

38 *Moral Treatise*, 11.4 (*Pr1*): "Narratur Genesis xi qualiter, cum turris Babel presumptuose erigeretur in celum, presumptuosis sublatum est vnitatis lingue commercium, et facta diuisione linguarum subsecuta est sectio populorum. Sic et philosophi dum ingenij sui machinas erexerunt in altum celestia et diuina

Some of the very authors in Peter's library, like Boethius of Dacia, have long been considered among the most important targets of the 1277 Condemnation. In such an atmosphere, Peter had to take great care in making the case, which Bacon had argued before him, that optics was essential knowledge for the moral improvement of Christianity.

Peter composed *The Moral Treatise on the Eye* to make that case in the period connected with and immediately following the 1277 Condemnation at the university in Paris, specifically between 1275/1276 and 1289, and probably towards the latter part of that range of years.[39] The work was transmitted earliest in Paris through the *pecia* system, which aimed to make it possible to acquire copies of recognized works made from sanctioned copy-texts at a much faster rate than had been possible before. Peter's work is included in the 1304 taxation list of *pecia* texts available at the de Sens family's *stationarius* shop.[40] As Louis-Jacques Bataillon has demonstrated, Peter of Limoges was a very well-established customer at this shop; a page with his account for copying work done there has sur-

curiose rimantes et de suis viribus presumentes, subtracta est eis vere et incommutabilis et indeficiens vnitas veritatis et ignorantie nebulis obuoluti et in reprobum sensum dati in sectas errorum varias sunt dispersi, et dicentes se esse sapientes, stulti facti sunt, Romanorum i."

39 On the dating of Peter's text, see Richard Newhauser, "Der 'Tractatus moralis de oculo' des Petrus von Limoges und seine *exempla*," in *Exempel und Exempelsammlungen*, ed. Walter Haug and Burghart Wachinger, Fortuna vitrea, 2 (Tübingen, 1991), 95–136.

40 Both the 1304 taxation list and the one from ca. 1275/1276 are found in the *Chartularium Universitatis Parisiensis*, ed. Denifle and Chatelain. For mention of the treatise in the earlier list by Guillelmus de Sens, see 1:644–50, no. 530; in the later list by Andreas de Sens, see 2,1:109, no. 642. On the work of the de Sens family and the *pecia* system, see Richard H. Rouse and Mary A. Rouse, *Manuscripts and Their Makers. Commercial Book Producers in Medieval Paris 1200–1500* (Turnhout, 2000), vol. 1, 73–97; Hugues V. Shooner, "La production du livre par la pecia," in *La production du livre universitaire au moyen âge: Exemplar et pecia, Actes du symposium tenu au Collegio San Bonaventura de Grottaferrata en mai, 1983*, ed. Louis-Jacques Bataillon et al. (Paris, 1988), 17–37, here pp. 23–24; Richard H. Rouse and Mary A. Rouse, "The Book Trade at the University of Paris, ca. 1250–ca. 1350," in *La production du livre universitaire*, ed. Bataillon et al., 41–114, here pp. 56–64 (with further bibliography on the *pecia* system); reprint in *Authentic Witnesses*, 259–338, here 282–94. See also Giovanna Murano, *Opere diffuse per exemplar e pecia*, Textes et études du moyen âge, 29 (Turnhout, 2005). For the pecia system in Italy, which preceded its growth in Paris, see Frank P.W. Soetermeer, *Utrumque ius in peciis. Aspetti della produzione libraria a Bologna fra due e trecento*, Orbis academicus: Saggi e documenti di storia delle università, 7 (Milan, 1997).

vived.[41] Peter's authorship of *The Moral Treatise on the Eye* is demonstrated in his own extant university papers, where he explicitly refers to the treatise four times and also gives a summary of some of his sources for chapter 11.[42] Further, his name appears in copies of the text throughout its history more often than that of any other person.[43] But often the transmission of the treatise was anonymous, so that early scribes and librarians were tempted to provide the text with an author they deemed likely to have composed it. In England, and later on the continent as well, the overwhelming choice for authorial attribution was Robert Grosseteste, due no doubt to Grosseteste's work on optics and as an analogy to the transmission of John of Wales' treatise *De lingua* which also circu-

41 Louis-Jacques Bataillon, "Comptes de Pierre de Limoges pour la copie de livres," in *La production du livre universitaire*, ed. Bataillon et al., 265–73.

42 Paris, Bibliothèque nationale de France MS. lat. 16390, fol. 13va: "Ad hoc autem, quod scolaris ad hanc sciencie perfeccionem attingat, tria principaliter requiruntur, uidelicet humilitas in intellectu, unde Salomon, Prouerbiorum xi: Vbi humilitas, ibi sapiencia. Et Mathei xi dicit Sapiencia increata: Confiteor tibi Pater etcetera; et asina Balaam etcetera, et Bernardus de scala Iacob, et Ysidorus Libro soliloquiorum et Ptolomeus in Almagesti: Qui inter sapientes humilior etcetera. Hec omnia require Libro de oculo, xi capitulo, de scolaribus, primo capitulello. Hec autem scolaris humilitas secundum Hugonem, Didascalicon libro tercio, tria continet documenta, uidelicet ut a nemine discere erubescat; ut nullam scienciam uel scripturam uilem habeat; ut cum scienciam adeptus fuerit, ceteros non contempnat. Propter primum ait Augustinus: Ego senex et tot annorum episcopus etcetera, et Paulus ab Anania et Glosa ibi. Hec omnia require ubi supra, scilicet Libro de oculo, xi capitulo de scolaribus, primo capitulello.... Secundo requiritur in uiro scolastico puritas in affectu. Sapiencia enim, cum sit speciosior sole, ut dicitur Sapiencie 7, mundum requirit hospicium et immundum habitaculum detestatur. Sicut enim candela madefacta etcetera, et oculus lippus etcetera. Hec omnia require in dicto Libro de oculo, predicto capitulo xi, primo capitulello.... Tercio requiritur studiositas in effectu – Bernardus Ad fratres de Monte Dei de cotidiana lectione, et Seneca: Cum multa percurreris, et idem: Qui scripta sua tenuerit. Set multi sunt scolares desides etcetera. Seneca: Desidioso studere etcetera. De talibus poeta: Sunt quidam, qui scire uolunt etcetera. Hec omnia require dicto Libro de oculo, xi capitulo, capitulello v°." The text has been printed in Louis-Jacques Bataillon, "Sermoni e orazioni d'ambiente universitario parigino nel sec. XIII," *Documenti e studi sulla tradizione filosofica medievale* 5 (1994): 297–329, here 307–11. See also Bériou, "Prédication," 108 n13.

43 See Spettmann, "Das Schriftchen," 318–19, and see the observation on this point, first mentioned by Spettmann, in Amédée Teetaert, "Jean Pecham," in *Dictionnaire de Théologie Catholique*, vol. 12/1 (1933), 100–40, here 114–15. For the attribution of *The Moral Treatise on the Eye* to Peter of Limoges in a codex of the thirteenth century, see for example, Bruges, Stadsbibliotheek MS. 167.

lated under the bishop of Lincoln's name.[44] Many late medieval German scribes, on the other hand, showed a preference for one of Peter's sources in the history of optics, namely John Pecham, under whose name the *editio princeps* appeared in Augsburg.[45] But the treatise on the eye was attributed to a number of other authors, as well, whether or not they were associated historically with an interest in optics: John of Limoges,[46] John of Wales,[47] Hugh of St. Victor,[48] Thomas Bradwardine,[49] Johannes Duns Scotus,[50] a Henricus or Henricus de Frimaria,[51] and, in one case, even Alhacen.[52]

The length of the chapters varies widely, some of the early ones and chapters 9 and 10 being only a page or two in length, while chapters 6, 7, 8, 11, and 12 fill many pages with expanded expositions and a large amount of *exempla*, verse, and other illustrative material that apply optical theory often to the moral critique of members of the church and the uni-

44 On the *De lingua* see Edwin D. Craun, *Lies, Slander, and Obscenity in Medieval English Literature: Pastoral Rhetoric and the Deviant Speaker*, Cambridge Studies in Medieval Literature, 31 (Cambridge, Engl., 1997), 18–19; Carla Casagrande and Silvana Vecchio, *I peccati della lingua. Disciplina ed etica della parola nella cultura medievale*, Bibliotheca Biographica. Sezione Storico-Antropologica (Rome, 1987), 141–74. For evidence of John of Wales as the author of this text, see Siegfried Wenzel, "The Continuing Life of William Peraldus's *Summa vitiorum*," in *Ad litteram. Authoritative Texts and Their Medieval Readers*, ed. Mark D. Jordan and Kent Emery, Jr. (Notre Dame and London, 1992), 135–63, here 142–43. For one of the earliest attributions of the *Tractatus moralis de oculo* to Grosseteste, see London, Lambeth Palace Library MS. 483. Many copies of Peter's work by English scribes contain no authorial attribution. For one such early manuscript that traveled as far as Estonia in the Middle Ages, see Richard Newhauser, Tiina Kala, and Meelis Friedenthal, "The Work of an English Scribe in a Manuscript in Estonia," *Scriptorium* 62,1 (2008): 139–48, plate 19; Meelis Friedenthal, *Tallinna Linnaarhiivi: Tractatus moralis de oculo*, Dissertationes Theologiae Universitatis Tartuensis, 13 (Tartu, 2008).

45 This is *Pr1*, used as the basis for this translation. On incunabula prints see George Sarton, *Introduction to the History of Science*, Publications of the Carnegie Institution of Washington, 376 (Baltimore, 1931), vol. 2/2, 1029–30. For an attribution of the *Tractatus moralis de oculo* to Pecham, see for example Basel, Universitätsbibliothek MS. B X 34.

46 See Paris, Bibliothèque Mazarine MS. 888 (1136).

47 See, for example, Assisi, Convento dei Frati Minori MS. Com. 442.

48 See Sankt Florian, Augustiner-Chorherrenstift MS. XI.164.

49 See, for example, Uppsala, Universitetsbibliothek MS. C 641.

50 See, for example, Trier, Stadtbibliothek MS. 574/1522 8°.

51 See Würzburg, Universitätsbibliothek MS. I.t.f.201.

52 See Sankt Florian, Augustiner-Chorherrenstift MS. XI.130.

versity. The following summary of the contents of the treatise emphasizes the discursive points made in each section:

> The first seven chapters lay out the procedure of the treatise (moving from what can be established *scientialiter* to what can be confirmed *moraliter*, i.e., through moral or spiritual interpretation, the transition often signaled by a phrase such as *modo consimili* [in a similar fashion], or the like). These chapters lay the foundations for the rest of the work by drawing heavily on Perspectivist optics:
>
> Chapter 1 establishes the seven parts of the eye: three coverings (uvea, cornea, sclerotic coat), three fluids (vitreous, crystalline [i.e., the pupil], albugineous), and one fine-textured membrane. These are related to the seven principal virtues.
>
> Chapter 2 explains that the softer components making up the eye are closer to the pupil and the harder ones further from it. This physical feature is related to the necessity, especially for prelates, to demonstrate both mercy and justice.
>
> Chapter 3 establishes the three types of physical vision: by direct lines of sight, by refracted lines, and by reflected lines. These are distinguished according to the veracity of the information they convey, and they are related to the types of spiritual vision.
>
> Chapter 4 asserts that both intromission and extramission, as well as the cooperation of the observer's will, are required for completed vision. These requirements are related to the necessity for the individual to cooperate willingly in the grace and spiritual power offered by God for the salvation of his/her soul.
>
> Chapter 5 makes note of the function of the internal common nerve, not the external eye, as the organ in which vision is completed, and this serves as the basis for the moral instruction to avoid rash judgment by looking below the surface of appearances.
>
> Chapter 6 explains thirteen optical illusions both physically and morally/spiritually: an observer located within fog cannot see the fog, but once removed from the fog the observer can see it (morally: a sinner does not see his own state of sin while in the midst of it); pushing against the eye with one's finger will make a visual object appear to be doubled (morally: an evil advisor, or bribes, can corrupt the just vision of a prelate or a prince); if at night an observer holds a dense object between his eye and a source of light and looks directly at the source of light, the object will appear to be doubled (morally: if a sinner looks at his sin in the light of God's word, both the sin and its punishment will be clear); a rod sticking out of water appears to be broken because of the magnifying quality of the water (morally: an upright person who is seen to enjoy physical pleasure

will be considered broken spiritually); an object barely visible to an observer can be seen more clearly if it is immersed in water (morally: carnal sins appear to be more important than spiritual sins); objects perceived in a mirror have a weaker image because reflection is weaker than direct vision (morally: seeing God's will directly means carrying out the intent of Scriptures; not carrying out God's will is like observing it in a mirror); an observer looking at the sun in a mirror placed in water sees three suns: one in the sky, one in the mirror, and one on the surface of the water (spiritually: contemplating God will reveal him to be threefold and singular); an observer at the center of a concave spherical mirror sees only himself because of the angles of reflection and incidence of that surface (spiritually: God knows and understands all things within himself); an object at the center of a concave spherical mirror cannot be seen by an observer located outside the center when the observer looks at the mirror (spiritually: we are unable to comprehend the divine essence in this life); shapes perceived in a mirror appear to be reversed (morally: the mirror of holy doctrine makes prosperity and misfortune, the proud and the humble, appear as the reverse of the way the world judges them); an object's size cannot be determined by refraction, i.e., when it is immersed in water (morally: sin can be comprehended only by someone who looks at it directly with the eye of reason); an observer perceives an object in water to be larger than it is, but if the observer is located in water and the object in the air, the object will appear smaller than it is because the density of the visual medium affects perception (morally: the perception of the rich by the poor, and vice versa, is conditioned by their social circumstances, but it is the poor who are great in divine sight); finally, the "moon illusion" is explained, that a heavenly body perceived near the horizon appears larger than when it rises into the sky (morally: the higher saintly people climb in worldly honors, the more they humble themselves).

Chapter 7 develops moral instruction on the basis of twelve natural properties of eyes: eyes come in pairs so they can strengthen each other (spiritually: nature commends the communal life); each eye is exactly like its companion (morally: our eyes should be similar in color, practicing what we teach, and in size, balancing appeals to the intellect and the affects); eyes are varied in color, and visual power changes with the eyes' color (spiritually: the worldly wise cannot see clearly in spiritual terms, but those who remember their mortality can see clearly); eyes are spherical, which is the simplest shape (spiritually: we should be singularly intent on eternal rewards); our eyes are covered by eyelids, but animals without eyelids have weak eyesight (morally: the vainglorious, who do not conceal their good deeds, will lose everything); eyes are located in the head (spiritually: the head is Jesus, to whom the intellect should be oriented); eyes are receptive to visible *species*, which is proven by staring through an opening of a house lit by bright sunlight and when one turns away one will see the image and shape of the opening (spiritually: the opening we should look at is the wound in Jesus' side on the cross); eyes are able to perceive another object, but not themselves (morally: it is foolish to judge others when one does not understand their thoughts); eyes are able to gain knowledge of visible things in different ways according to variations in their distance (morally: sinners see God and heavenly reward only from a far distance); eyes are able to establish the direction for the entire body, passing information to the rest of the body freely (morally: we should be eager to share liberally the good qualities that we have); eyes give an indication of the secrets of the mind, which is revealed by eye movement and what one looks at (morally: one should persevere in what one undertakes, which is equivalent to moderate eye movement, and should not watch obscene and shameless spectacles); eyes are drowsy after being awake for a long time (morally: we should redirect ourselves from external matters to internal ones).

The next chapters develop sets of moral instruction using cultural conceptions surrounding the eyes, in particular guidelines for specific audiences (especially students and prelates):

Chapter 8 develops the conception of types of eyes that characterize those sinning by the seven capital vices. The proud have arrogant and haughty eyes that blind the eye of their reason; the envious have squinting eyes and cannot see the good fortune of their fellow human beings, but do see that person's misfortune; the wrathful have agitated eyes that can infect others if they look at them in a state of great anger; the slothful have eyes that are sluggish and sleepy; the eyes of the avaricious have been blinded by the dust of worldly possessions; the gluttonous are said to have fleshly eyes that are sharpest when they see food; and the lustful have eyes that are attracted to and manipulated by women, but also all of their senses are susceptible to the allure of women.

Chapter 9 demonstrates that the eyes are responsible for more expenses than any other bodily member.

Chapter 10 counsels to suffer the loss of one's eyesight with equanimity.

Chapter 11 applies seven conditions necessary for vision to occur to the instruction of students. These conditions are: the required disposition of the visual organ, which is adversely affected by swelling, dust, and fluid – this demonstrates that a student must avoid pride, avarice, and lust by being humble, unaffected by wealth, and chaste; the presence of an object – students must commit the content of their books to memory, not use them as objects for show; a proportionate amount of distance – students must maintain objectivity in order to learn correctly; the density of the visual object – just as air cannot be seen, types of (secular) knowledge without substance and not based on Scriptures are empty and vain; the application of the visual faculty – students must turn their active attention to what they are studying; a duration of time – students should dwell on their reading and not rush through it quickly; and the diffusion of light illuminating the medium – in order for the spiritual eye to see, a divine light is required.

Chapter 12 develops instruction for prelates based on seven properties of the eyes: they are located in a very prominent part of the body – prelates are taught to surpass others in the holiness of their lives; eyes keep watch for the entire body – a prelate must serve as a watchman for the church; the eye contemplates things that are distant from it, not those very close to it – prelates are taught to scrutinize the behavior of those close to them; eyes become frail and weakened in an aged body – the prelates and clerics of the modern church have grown lax and unethical; eyes weep as if showing sympathy with the pains in other bodily members – a prelate should shed tears of compassion for the sins of the congregation; eyes have eyelids with hairs, the size of which does not increase – prelates are taught that their temporal possessions should not increase over time; and an injury that in other bodily members is thought to be moderate is considered serious when it happens to the eyes – the same sin is greater in a prelate and a priest than in a member of the laity.

After having explained the scientific and moral foundations of vision, in the final section of the treatise Peter focuses on spiritual sight:

Chapter 13 presents four things on which one's spiritual eyes should be focused: the filthiness of one's own iniquity, the hardship of another person's misfortune, the bitterness of everlasting punishment, and the recompense of a blessed life.

Chapter 14 interprets in spiritual terms three things that delight the corporeal eyes: the translucence of water (i.e., the Virgin Mary), the beauty of flowers and plants (i.e., saints' exemplary deeds), and the radiance of a mirror (i.e., Jesus).

> Chapter 15 presents seven properties of divine sight: God's eye contemplates all things, it is awe-inspiring, it produces tears of penitence, it promotes hard work, it contributes to spiritual strength, it heals the infirmity of the soul, and it leads to the kingdom of glory.[53]

The macro-structure of the treatise recapitulates the procedure established in the early chapters, moving from physical vision to spiritual sight via the process of moral interpretation. First Peter establishes a foundation made up of the new teachings on Perspectivist optics, and then he draws ethical and spiritual implications important for his text, to the point of allegorizing the scientific explanations for vision themselves,[54] while at the end of the text the quality of spiritual sight becomes the focus of the work.

This skeletal outline of the contents of *The Moral Treatise on the Eye* does not indicate the degree to which these discursive elements are given substance by a large number of illustrative passages in verse, images, metaphors, and above all *exempla*. It was, in fact, as a member of the genre of *exempla*-collections that Peter's text was most often transmitted in the Middle Ages.[55] As such, it served as a storehouse of information on optics that made the scientific discourse of Roger Bacon or Alhacen fit for use in the pulpit before a much wider audience and one less educated in natural philosophy and optics.[56] The treatise became very popular and is still extant or can be attested in over 260 copies.[57] It was used often to

53 See also Charles Trice Martin, "Preface," in *Registrum epistolarum fratris Johannis Peckham, archiepiscopi cantuariensis. ... Published by the authority of the lords commissioners of Her Majesty's Treasury, under the direction of the master of the rolls*, ed. Charles Trice Martin, Rerum Britannicarum Medii Aevi scriptores, or Chronicles and memorials of Great Britain and Ireland during the Middle Ages, 77 (London, 1885; reprint [Nendeln], 1965), vol. 3, xxvii–cxlv, here lxxxi–xcvii, which contain a description of the contents of *The Moral Treatise on the Eye* that Martin believed at the time might have been the work of Pecham.

54 Dallas G. Denery, II, *Seeing and Being Seen in the Later Medieval World. Optics, Theology and Religious Life*, Cambridge Studies in Medieval Life and Thought, 4th ser., 63 (Cambridge, Engl., 2005), 78.

55 See Richard Newhauser, "Nature's Moral Eye: Peter of Limoges' *Tractatus moralis de oculo*," in *Man and Nature in the Middle Ages*, ed. Susan J. Ridyard and Robert G. Benson, Sewanee Mediaeval Studies, 6 (Sewanee, TN, 1995), 125–36.

56 See Suzannah Biernoff, *Sight and Embodiment in the Middle Ages* (Houndmills and New York, 2002), 122–24; David L. Clark, "Optics for Preachers: The *De oculo morali* by Peter of Limoges," *Michigan Academician* 9 (1977): 329–43.

57 For lists of manuscript copies, see Schleusner-Eichholz, "Naturwissenschaft," 258–60, supplemented by Newhauser, "Nature's Moral Eye," 133, and most

compose sermons. In England, to give only a few examples from a single country, the Benedictine Robert Rypon (d. after 1419) quotes the treatise in two sermons, as do a number of other preachers of the fifteenth century whose texts are preserved in both Latin and English.[58] But as was noted at the outset, Rypon and the other preachers who drew on the text in England quoted the treatise as the work of Robert Grosseteste, not Peter of Limoges.

Despite its popularity, only one complete translation of Peter's work has appeared before now, a fifteenth-century rendering into Italian, namely:

> *Libro de l'occhio morale et spirituale uulgare. Epistola interpretis Theophili Romani, O. Erem. S. Aug., ad F. Dominicum da Ponzo tanquam auctorem, qui vero in breui praeloquio dicit, se haec invenisse in un Libretto. Sub finem verus auctor dicitur reverendo professor di sacra theologia Maestro P. Lacepiera*. Trans. Teofilo Romano. Venetiis: Joannes Rubeus Vercellensis, 1496.[59]

This translation helps document the way in which Peter's work "had some vogue in Italy in the late fifteenth century."[60] No critical edition of *The Moral Treatise on the Eye* exists yet,[61] and nevertheless a sufficient interest in Peter's work has developed, especially in the past two years, to create something of a demand for the translation of the *Treatise* now. To meet that demand before a critical edition is available, I have based the present translation on the earliest printed text, copies of which are extant in a number of libraries around the world: *Johannis Pithsani archiepiscopi*

---

recently Richard Newhauser and István Bejczy, *A Supplement to Morton W. Bloomfield et al., 'Incipits of Latin Works on the Virtues and Vices, 1100–1500 A.D.'*, Instrumenta Patristica et Mediaevalia, Research on the Inheritance of Early and Medieval Christianity, 50 (Turnhout, 2008), no. 5532.

58 Siegfried Wenzel, *Latin Sermon Collections from Later Medieval England: Orthodox Preaching in the Age of Wyclif*, Cambridge Studies in Medieval Literature, 53 (Cambridge, Engl., 2005), 328. See also the English sermon preserved in Oxford, Bodleian Library MS. Laud misc. 706, fols. 144r–152v, edited by Patrick J. Horner, "An Edition of Five Medieval Sermons from MS. Laud misc. 706," Ph.D. Dissertation, State University of New York at Albany, 1975, 111–46.

59 Sarton, *Introduction to the History of Science*, vol. 2/2, 1029–30; Ludwig Hain, *Repertorium bibliographicum* (Stuttgart, 1831; reprint Berlin, 1925), vol. 2/1, 224, no. 9805.

60 Michael Baxandall, *Painting and Experience in Fifteenth Century Italy. A Primer in the Social History of Pictorial Style*, 2nd ed. (Oxford and New York, 1988), 104.

61 I am currently completing the first critical edition of Peter's work.

Canthuariensis ordinis fratrum minorum liber de oculo morali foeliciter incipit (Augsburg: A. Sorg, 1475 [?]) [*Pr1*]. A comparison of this text with one of the better manuscripts of the *pecia* transmission (Paris, Bibliothèque nationale de Paris MS. lat. 16395 [*P7*]) and one of the manuscripts of the work from Peter's own library (Paris, Bibliothèque nationale de Paris MS. lat. 16396 [*P8*]) reveals that *Pr1* closely follows the arrangement of material in the treatise as this is represented in its authoritative transmission by *pecia*. Besides showing an expected range of variations in wording, it adds relatively few sentences to what is found in the text of the *pecia* transmission, but it frequently alters sub-titles in the chapters of the treatise as transmitted by *pecia*. The noteworthy lengthier additions, deletions, or alterations found in *Pr1* in comparison with the authoritative transmission of the text in manuscript are as follows:

> A title and colophon attributing the work to John Pecham have been added. Headings are added, contracted, or expanded throughout the treatise as titles for sub-sections of the chapters. In chapter 2, a short sentence has been added (see p. 8); in chapter 7.10, two sentences have been added to a quotation from one of Seneca's letters (p. 70); in chapter 8.1, three sentences have been added, taken from what is represented as Dindimus' first letter to Alexander the Great (p. 78); an eye-skip has deleted needed material in chapter 8.5 (p. 94); and one verse has been added to the description of gluttony in chapter 8.6 (p. 96). Finally, in chapter 11.5, three verses have been added from John of Salisbury's *Policraticus* (p. 140), and one brief sentence has been added to chapter 11.7 (p. 144).

In the translation, I have followed *Pr1* wherever possible, that is to say wherever it makes grammatical sense and does not unrecognizably alter the content of a given passage. Where such substantive alterations occur, I have emended the text that is rendered in English by borrowing from *P7P8*, a designation that indicates how close these manuscripts are to each other (or from either *P7* or *P8* in cases where they disagree with each other). I have marked emended portions of the translation with an asterisk (*). In a few instances, I have introduced editorial emendations when all three sources give a faulty reading. Other material in *Pr1* that does not amount to a grammatical rupture of the text, but that still varies significantly from the transmission of Peter's text in the best manuscripts, has likewise been marked in the translation with an asterisk. These latter passages amount most often to additional subheadings or the few added sentences within the chapters. The reader will find the Latin variants of *P7*, *P8*, and *Pr1* for all passages marked with an asterisk in the section of Emended Readings and Significant Variations at the end of the translation. Typographical errors and minor grammatical or orthographic variations in *Pr1* have been translated in their corrected or normalized forms and are not commented on further. These variants include the following: chapter 2: *sedat* for *sedant*; chapter 6.5: *maioria* for *maioris*, *abscondtie* for *abscondite*; chapter 7.2: *quiidam* for *quidam*; chapter 7.6: *apperuit* for *aperuit*; chapter 8.4: *Palmiste* for *Psalmiste*, *diuitum diuitum* for *diuitum diuites*; chapter 8.5: *appetit dignitatem* for *appetit dignitate*; chapter 8.7: *muliere ... preparatus* for *muliere ... preparata*, *cenit* for *cernit*; chapter 11.1: *sermotionatio* for *sermocinatio*, *pulle* for *puelle*; chapter 11.5: *premisque* for *premiisque*; chapter 11.7: *pandetur* for *pandentur*; chapter 12.2: *inutitur* for *innititur*; chapter 12.3: *domo suo* for *domo sua*, *dignes* for *dignos*, *texicatus* for *toxicatus*; chapter 12.4: *vemora* for *femora*; chapter 12.7: *subditie*, *xemplo* for *subditi exemplo*; chapter 13.1: *apparente delectatione* for *apparenti delectatione*; chapter 13.2: *Vn-* (at the end of a line) for *Vnde*; chapter 13.3: *contremescant* for *contremiscant*; chapter 14.3: *maculas deprehendendos* for *maculas deprehendendas*.

As the reader will observe, the notes to the text are more extensive than those in other volumes in the series Mediaeval Sources in Translation. The need for a wider-ranging apparatus has been occasioned partially by the technical characteristics of optics that serve as the foundation of the *Treatise*. Thus, the footnotes here not only identify Peter's sources and a number of analogues to his text, but also comment on the way Peter uses, and at times reinterprets, the scientific principles he adopted from other works centered on Perspectivist optics, especially those by Alhacen, Roger Bacon, and John Pecham. To make the text as accessible as possible to modern readers, the notes also flesh out the connotations and implications of Peter's use of all of his sources: scientific, historical, narrative, and theological. In a number of cases, I have pointed out further readings on topics essential to Peter's composition of the *Treatise*, especially in the moral tradition of the Middle Ages. For students who do not yet have a sufficient command of Latin but wish to follow up on the material Peter used to compose the *Treatise*, translations into English or French of the sources and analogues to passages actually quoted or alluded to in his text will be found in the bibliography. Each translation is listed in the bibliography immediately following the edition of the original text I have made use of. Wherever they were available, I have included translations of complete works here, though sometimes only translated excerpts of material adopted by Peter exist and, of course, it was not always possible to find translations based on the best available editions of the Latin works. Special note has been made of translations accessible online at the present moment.

My translation aims for "translation equivalence," when that technical term can be defined as a matter of textual and practical equivalence, in particular.[62] I have attempted to make the English rendering an equivalent of the Latin text's information and cohesion (textual equivalence), but I have also attempted to render the implicatures of Peter's text (practical equivalence). Thus, the translation attempts to be close to the communication of information in the original without being slavishly literal. It also attempts to reproduce the stylistic registers of the original: where Peter is technical in the explanation of optics, the translation attempts to be so as well; where Peter represents direct speech, the translation attempts to be conversational in tone. The translations of all texts quoted by Peter are my own, including previously translated material, such as the Bible. The forms of biblical names used in the translation are adopted from those found in *The New English Bible with the Apocrypha*.[63] Finally, where *Pr1* uses a word to indicate a number, I have also; where a numeral appears in the Latin, a numeral appears in my translation.

Tempe, Arizona; August, 2011

62 See especially Dorothy Kenny, "Equivalence," in *Routledge Encyclopedia of Translation Studies*, 2nd ed., ed. Mona Baker and Gabriela Saldanha (London and New York, 2009), 96–99; Mona Baker, *In Other Words: A Coursebook on Translation*, 2nd ed. (London, 2011).

63 *The New English Bible with the Apocrypha* (New York, 1971).

PETER OF LIMOGES

The Moral Treatise on the Eye

Here Begins the Moral Treatise on the Eye*

If we want to contemplate the law of the Lord[1] diligently, we are going to recognize very easily that those matters which pertain to vision and the eye are referred to more frequently than any others in holy writings. From this fact it is obvious that a consideration of the eye and of the things that are related to it is very useful in order to gain a more complete understanding of divine wisdom.[2] Thus, as I am about to say a few words on the eye – since the fortifying of souls is contained in them –, I want first of all to compose a brief section dealing with the matter scientifically and, second, after this to treat it morally, insofar as "he who formed the eye,"[3] "whose eyes are turned to the pauper,"[4] will consider it worthwhile to illuminate the blind eye of my reason.

1 Cf. Ps 1:2.

2 Cf. Roger Bacon, *Perspectiva*, 3.3.1, ed. and trans. David C. Lindberg, in *Roger Bacon and the Origins of Perspectiva in the Middle Ages: A Critical Edition and English Translation of Bacon's Perspectiva, with Introduction and Notes* (Oxford and New York, 1996), 322, which is adopted partially verbatim by Peter. For parallels between the two texts, see Richard Newhauser, "*Inter scientiam et populum*: Roger Bacon, Peter of Limoges, and the 'Tractatus moralis de oculo,'" in *Nach der Verurteilung von 1277. Philosophie und Theologie an der Universität von Paris im letzten Viertel des 13. Jahrhunderts. Studien und Texte / After the Condemnation of 1277. Philosophy and Theology at the University of Paris in the Last Quarter of the Thirteenth Century. Studies and Texts*, ed. Jan A. Aertsen, Kent Emery, Jr., and Andreas Speer, Miscellanea Mediaevalia, 28 (Berlin and New York, 2001), 682–703, here 700 (appendix, part A).

3 Ps 93:9. All citations from the Vulgate Bible are based on *Biblia sacra iuxta Vulgatam Clementinam. Nova editio*, ed. Alberto Colunga and Laurentio Turrado, 4th ed. (Madrid, 1965). Translations are my own.

4 Ps 10:5.

Thus, the present work is divided into chapters, fifteen in number:

The first, on the number of parts comprising the eye.
The second, on the order of the parts in the composition of the eye.
The third, on the number of types of vision.
The fourth, on the means of vision.
The fifth, on the organ in which vision is completed.
The sixth, on thirteen extraordinary phenomena that contain moral lessons on eyesight.
The seventh, on moral instruction according to twelve properties that have been ascertained in the corporeal eye.
The eighth, on seven differentiations of eyes according to the differentiation of the seven capital vices.
The ninth, on the costliness of the eye in respect to the other parts of the human body.
The tenth, on how one must suffer the loss of the corporeal eyes with equanimity.
The eleventh, on lessons for students based on the seven conditions that are necessary for sight.
The twelfth, on the instruction of prelates based on seven properties of the eyes.
The thirteenth, on four things that the spiritual eyes ought to contemplate continually.
The fourteenth, on three visible things that delight the eye.
The fifteenth, on the sevenfold eye of the divine gaze.

CHAPTER ONE

On the Number of Parts Comprising the Eye

It ought to be known, therefore, that there are three coverings in the composition of the eye, as can be understood from the opinion of Constantine in the book that he composed *On the Eye*,[1] and from the words of Alhacen* in the first book of *Perspectivist Science*.[2] The first of these coverings is said to be the "uvea," seeing that it is like a grape (*uva*); the second, the "cornea," since it is shiny and glossy like horn (*cornu*); the third is called "sclerotic"* or "consolidative," since it is solid or hard. The eye also has three fluids: The first takes the name "vitreous," since it bears the likeness of liquefied glass (*vitrum liquatum*); the second is said to be "glacial" or "crystalline," since it imitates crystal or ice; the third is called "albugineous," since it is very similar to the white (*albugo*) of an egg. In addition, the eye has a fine-textured and small membrane containing the three fluids that have just been touched on which is similar to a spider's web, and because of this it is named the "spider web." Lastly, the eye has eyelids which cover it and protect it from external injuries.

The crystalline fluid located in the midst of the other fluids is called the "pupil," and the visual power exists in it as its subject, though not in a primary sense, because the common nerve is the source organ in which the two optical nerves coming from the eye intersect in the manner of a cross, and in that common nerve the visual power has its seminal source,

1 Cf. Constantine the African, *Liber de oculis*, 1.2.3–1.3.6, ed. Pierre Pansier, in *Liber de oculis, Galieni littere ad Corisium de morbis oculorum et eorum curis*, Collectio ophtalmologica veterum auctorum, 7 (Paris, 1909–1933), 168–70.

2 Cf. Alhacen, *De aspectibus*, 1.6 [5.1–38], ed. and trans. A. Mark Smith, in *Alhacen's Theory of Visual Perception. A Critical Edition, with English Translation and Commentary, of the First Three Books of Alhacen's De Aspectibus, the Medieval Latin Version of Ibn al-Haytham's Kitab al-Manazir*, Transactions of the American Philosophical Society, 91, Parts 4–5 (Philadelphia, 2001), vol. 1, 11–22 (Latin text), vol. 2, 348–55 (English translation).

and from there it is diverted into the pupil of each eye.[3] Hence, it is clear what an elegant simile the prophet David used when he said, "Protect us, Lord, like the pupil of your eye."[4] For as was mentioned already, the crystalline or glacial fluid is called the "pupil," which is enclosed by two fluids and one membrane and three coverings and also, beyond that, by the eyelids.[5] Nevertheless, it experiences the continuous influx of spirits and powers from the original bountifulness of visual power that exists, as was briefly mentioned, in the cruciform section of the nerves. From all of this it is clearly apparent that seven things are required for the protection of the pupil.

That seven things are necessary to protect the spiritual pupil.*

In a similar way, for the perfect protection of the spiritual pupil, that is of the soul, we require seven things, namely, the seven principal virtues, i.e., the three theological and the four cardinal. And on top of that ought to be added the continual influx of the gifts of grace pouring out from the bountifulness of the one who was crucified. Through these our spiritual pupil is to be protected by the Lord.[6] Therefore, the most wise David prayed fittingly when he said, "Protect me, Lord," etcetera.[7] Moses, the holy man taught by the same spirit, used the same simile when he said in Deuteronomy 32, "He encircled him and taught him and protected him like the pupil of his eye."[8] By the fact that nature has fortified the eye with such protection we are taught spiritually to carefully protect the eyes, i.e., lest having been negligent we are forced to say at last, "My strength has

3 Cf. Roger Bacon, *Persp.*, 1.2.3, ed. and trans. Lindberg, 32. Note that Peter has changed his source here, for Bacon writes that the visual power exists in the pupil as its subject since it is the first thing altered (i.e., by the visual *species* of the perceived object), but it is not the source of the visual power since the common nerve is the source organ ("Et in [pupilla] est virtus visiva sicut in subiecto, quod primo immutatur, licet non radicali, quoniam nervus communis est organum radicale ... ."). For verbal parallels between Bacon's text and Peter's adoption of it here, see Newhauser, "*Inter scientiam et populum*," 700–01 (appendix, part B).

4 Cf. Ps 16:8.

5 Note that in the next section these protective coverings of the eye are referred to as the "seven guardians of the pupil."

6 Cf. Roger Bacon, *Persp.*, 3.3.1, ed. and trans. Lindberg, 322. For close verbal parallels between Bacon's text and Peter's adoption of it here, see Newhauser, "*Inter scientiam et populum*," 701 (appendix, part B).

7 Ps 16:8.

8 Deut 32:10.

abandoned me; and the light of my eyes – it, too, is not with me."[9] Since the spiritual pupil, i.e. the soul, is to be protected by seven virtues as by seven guardians of the pupil, as was mentioned already, it is no wonder if someone who has been abandoned by his virtue has lost the light of his eyes. And therefore in Proverbs 7 the Lord, too, wishing his law to be kept very strictly, as one who has ordered that his commands be closely followed, explicitly compared the keeping of his law to the protection of the pupil: "Keep my law like the pupil of your eye."[10]

9 Ps 37:11. Note that *virtus*, which occurs in the biblical verse meaning "strength," is taken here to refer to the seven virtues guarding the soul, mentioned in the next sentence.

10 Prov 7:2.

CHAPTER TWO

On the Order of the Parts in the Composition of the Eye

Concerning the coverings of the eyes that were previously mentioned, one should bear in mind that while the coverings which come together to protect the pupil are distinguishable in their softness and hardness, still, the softer covering is closer to the pupil, while the harder one is farther away from it.

*On the mercy and justice of prelates**.

From this fact the eyes of the church, that is to say prelates, can take away a spiritual lesson, namely that just as they exhibit the gentleness of mercy, so they must not omit the punishment of justice. For as Gregory says, "Justice is greatly flawed without mercy, if one should be practiced without the other."[1] Nevertheless, prelates will be more prone to mercy so that mercy will be closer to them. Let justice be added to this, an example of which prelates may observe in Elijah. Indeed, Elijah stood out as if he were a type of eye that looked at things from afar, so to speak, and foresaw future events. "For he who is now called a prophet formerly used to be called a seer," as it is said in First Samuel 9.[2] This eye was composed most correctly, for it was more prone to mercy than to justice, and so it had mercy to the degree that it observed the other quality. Concerning Elijah it is said in Second Kings 2: "My father, my father, the chariot of Israel and its horseman!"[3] For a prelate should be a chariot, first and foremost, supporting the weak through mercy. He should be a horseman by goading the rebellious through the severity of justice. And it is, thus, fitting that the chariot is spoken of first and then the horseman*: Elijah observed this order and by his example taught prelates to observe this same order. In

1 Gregory the Great, *Moralia in Iob*, 20.5.14, ed. Marc Adriaen, CCSL 143A (Turnhout, 1979), 1012; idem, *Registrum epistularum libri i–vii*, 1.24, ed. Dag Norberg, CCSL 140 (Turnhout, 1982), 30; idem, *Regulae pastoralis liber*, 2.6, PL 77:38A.

2 1Sam 9:9.

3 2Kgs 2:12.

Luke 10,[4] the Samaritan, which means "guardian"[5] (by whom prelates are signified), poured oil and wine on the wounds of the man who had been beaten, and in a fitting order: first the oil, in which the gentleness of mercy is represented, and second the wine, in which the severity of discipline is seen.

Princes and prelates who are more prone to severity than mercy or leniency should hear and give heed to what the most prudent Helinand has to say about this:[6] "Prelates," he says, "and princes should, like physicians, never use harsh remedies unless they despair of achieving the desired state of health through the benefit of mild remedies. For this reason, Lucius says, 'It is fitting that a prelate or a prince be a man of advanced years in his morals, and one who should formulate more moderate advice,'[7] and he should put himself in the place of physicians who cure diseases (sometimes by purging those who have eaten too much, sometimes by feeding those who are undernourished) and who soothe pains (sometimes with a cautery, sometimes with a warm poultice). For if lute players and other minstrels concentrate with great diligence on how to curb the imperfection of one string that is out of tune and make it harmonious again with the others, with how much solicitude should the prelate or the prince govern in order to make his subjects once more like those who are 'harmonious and live together in a house under one rule'?[8] It is clearly safer to make the strings looser rather than to stretch them tighter. For the tension* of slack strings can be corrected by the effort of a craftsman, but once a string has been broken it cannot be repaired by any craftsman. For this reason:

4 Cf. Luke 10:34.

5 Jerome, *Liber interpretationis Hebraicorum nominum*, Luc., ed. Paul de Lagarde, CCSL 72 (Turnhout, 1959), 142.

6 For the background of the long citation that begins here and extends to the end of the material taken from Claudian, see below at n11.

7 Cf. John of Salisbury, *Policraticus*, 4.8, in *Policraticus I–IV*, ed. Katharine S.B. Keats-Rohan, CCCM 118 (Turnhout, 1993), 259. The attribution of this statement to a "Lucius" in the *Policraticus* seems to depend on the elision of the first two elements in the text on which this passage is based: Sextus Julius Frontinus, *Strategemata*, 4.7.3, ed. Robert I. Ireland (Leipzig, 1990), 107, which begins: "L. Paulus imperatorem senem moribus dicebat esse oportere … ." See below, n11. Cf. Frederic C. Tubach, *Index exemplorum. A Handbook of Medieval Religious Tales*, FF Communications, 204 (Helsinki, 1969; 2nd ed. 1981), no. 3952.

8 Ps 67:7.

Let the prince be slow in punishing, in rewarding, quick;
And let him grieve whenever he is compelled to be harsh.[9]

Of these matters the one deals with mercy, the other leniency, concerning which Solomon says: 'Mercy and truth will not forsake you. Bind them about your neck and write them on the tablets of your heart and you will find good grace and discipline in the sight of God and human beings.'[10] For grace is owed to mercy; to justice, discipline. Whoever is without the first is hateful to all; whoever is without the second is contemptible to all. Concerning this, Claudian also says:

You should be pious above all; for although we are surpassed in every
Virtue, mercy alone makes us equal to the gods.
Nor should you do what you are at liberty to do, but what will be fitting* for you
To have done, and let a regard for honorable behavior take over your mind."[11]

Chrysostom says on Matthew: "If we err by imposing a moderate penance, is it not better to account for this action as due to mercy than to resort to punishment because of cruelty?"[12] If modern prelates were like this, they would be similar to the corporeal eye, and then they would be loved by their charges as their eyes – and even more than their eyes.

9 Ovid, *Ex Ponto libri quattuor*, 1.2.121–22, ed. John A. Richmond (Leipzig, 1990), 8. Cf. Hans Walther, *Proverbia sententiaeque latinitatis medii aevi*, Carmina medii aevi posterioris latina, 2/5 (Göttingen, 1967), no. 29816.

10 Prov 3:3–4.

11 Claudius Claudianus, *Panegyricus dictus Honorio Augusto quartum consuli*, 267–68, 276–77, ed. John Barrie Hall, in *Carmina* (Leipzig, 1985), 71. Cf. Walther, *Proverbia*, no. 29757a (for the first two verses) and no. 18594 (for the final two verses). The entire passage beginning with the reference to Helinand is taken ultimately from Helinand of Froidmont, excerpted in the *Flores a Vincentio Bellovacensi collecti*, 1.15, PL 212:737A–D. Peter probably knew the work of Helinand only through excerpts, such as in the large *florilegium* by Vincent of Beauvais. For a text closer to Peter's quotation than what is found in the PL text of Helinand's work, see Vincent of Beauvais, *Speculum doctrinale*, 7.18, in *Speculum quadruplex; sive Speculum maius* (Douai, 1624; reprint Graz, 1965), vol. 2, 569b–d. Some of the same material, including the quotation from "Lucius," as observed above in n7, is also found in John of Salisbury, *Policr.*, 4.8, in *Policraticus I–IV*, ed. Keats-Rohan, CCCM 118:259–61.

12 Cf. Hrabanus Maurus, *Expositio in Matthaeum*, 7.23.4, ed. Bengt Löfstedt, CCCM 174A (Turnhout, 2000), 592 (PL 107:1066A) (and cf. John Chrysostom, *In Matthaeum homiliae xc*, 72.1, PG 58:668).

And thus Paul acted in this way, who was most merciful toward his charges when he said, "If someone is infirm, am I not also infirm? If someone is led astray to evil, am I not also pained?" – 2 Corinthians 11.[13] And again in Galatians 4: "You are my children, with whom I am in labor again."[14] And in Philippians, chapter one: "God is my witness how much I yearn for you all with the affection of Jesus Christ."[15] For that reason he was loved by his charges more than their eye. For he says in Galatians 4: "You received me as if I were an angel of God. For I bear witness for you that if it had been possible, you would have torn out your eyes and given them to me."[16] But today because of the hard-heartedness of prelates, the opposite often happens, for sometimes the charges wish to tear out the eyes of their prelate.

13 2Cor 11:29.

14 Gal 4:19.

15 Phil 1:8.

16 Gal 4:14–15.

CHAPTER THREE

On the Number of Types of Vision

Perspectivist authors make a distinction between three types of vision with the eye. The first is by means of straight lines, the second by means of refracted lines, and the third by means of reflected ones. Of these, the first is more perfect than the others, the second more certain than the third, and the third less certain.[1]

*On three types of vision among human beings**.

In the same way, spiritually speaking, we can describe three types of vision among human beings: The first is perfect, which will exist in the state of glory after the final resurrection. The second is in the soul separated from the body and contemplating the divine essence in the highest heaven until the resurrection, and this vision is weaker than the first. The third is in this life, which is the weakest of all, and it has to take place through reflection, just as the vision by which something is seen in a mirror also has to take place through the intermediary of reflected lines. For this reason, it is also called a mirror-like vision by the apostle Paul. "Now," he says, "we see through a mirror and dimly, but at the time of glory, face to face,"[2] and after the resurrection we will see with full directness, but before that only at an oblique angle to that directness or fullness, because the soul will not be filled by the directness or fullness of vision before it has been united with its body. For, as Augustine says, among the souls in heaven already blessed by the sight of God there exists a certain "natural longing to restore their bodies to

1 Cf. Alhacen, *De aspect.*, 4.1 [1.1–2], ed. and trans. A. Mark Smith, in *Alhacen on the Principles of Reflection. A Critical Edition, with English Translation and Commentary, of Books 4 and 5 of Alhacen's De Aspectibus, the Medieval Latin Version of Ibn al-Haytham's Kitab al-Manazir*, Transactions of the American Philosophical Society, 96, Parts 2–3 (Philadelphia, 2006), vol. 1, 3 (Latin text), vol. 2, 295 (English translation).

2 1Cor 13:12.

life," from which they are held back lest they be brought to God in their totality.[3]

From this it is clear how suitably the wisest Perspectivist spoke, the apostle Paul, when he called the vision of heaven face to face, but the vision of the journey there mirror-like and only partial, First Corinthians 13: "We see now through a mirror and dimly; then, however, it will be face to face. Now I know partially; then I will know just as I, too, have been known."[4] And just as someone who desires to see his best friend after a long time would not be satisfied if he could see him only in a mirror and could not also see him in person, so in this present moment, although we see God in a mirror, "the eye is not satisfied with seeing," Ecclesiastes 1.[5] But in the future, when we will see God by direct vision, and face to face, that verse by the Psalmist will be fulfilled: "I will be satisfied when your glory appears."[6] And just as the apostles "rejoiced when they had seen the Lord," John 20,[7] so too will we rejoice when we will see God with our eye, insofar as we will have been made Perspectivists and investigators in glory by then. We will delight ineffably in the greatest good, which we will behold, and with inexpressible joy we will break out with the sound of praise, according to the verse in Isaiah 52: "The voice of your watchmen: they have lifted up their voice; together they will praise, since they will see eye to eye."[8]

3 Augustine of Hippo, *De Genesi ad litteram libri duodecim*, 12.35, ed. Joseph Zycha, CSEL 28.1 (Vienna, 1894), 432, where Augustine's point is that the disembodied souls in heaven have a natural longing to manage their bodies that holds them back from going on in full force to the highest heaven, something they will be able to do only after they are reunited with their spiritual (no longer merely corporeal) bodies at the time of the resurrection. For the first two paragraphs, cf. Roger Bacon, *Persp.*, 3.3.2, ed. and trans. Lindberg, 328. The close verbal parallels between Bacon's text and Peter's are noted in Newhauser, "*Inter scientiam et populum*," 701–02 (appendix, part C).

4 1Cor 13:12.

5 Qoh 1:8.

6 Ps 16:15.

7 John 20:20.

8 Isa 52:8.

CHAPTER FOUR

On the Means of Vision

Further, as Ptolemy proves in his *Optics*, not only is the intromission of the form of the visible object required for vision, but also the extramission and cooperation of one's own *species* and power.[1]

In the same way, spiritual vision requires not only that the soul receive powers and grace from without, that is to say from God, but that it cooperate with them internally through its own power.

1 Ptolemy is, in fact, an exponent of the extramission theory of vision, which accounted for vision by postulating a visual flux that emanates from the eye. When the flux touches an object, vision occurs. Sight, then, is closely related to touch in its operation for Ptolemy. Cf. Ptolemy, *Optica*, 2.13, ed. and trans. Albert Lejeune, in *L'Optique de Claude Ptolémée dans la version latine d'après l'arabe de l'émir Eugène de Sicile. Édition critique et exégétique augmentée d'une traduction française et de compléments*, 2nd ed., Collection des travaux de l'Académie Internationale d'Histoire des Sciences, 31 (Leiden, 1989), 17; *Ptolemy's Theory of Visual Perception: An English Translation of the Optics with Introduction and Commentary*, trans. A. Mark Smith, Transactions of the American Philosophical Society, 86/2 (Philadelphia, 1996), 74–75. For a combination of the extramissionist and intromissionist positions as found in Plato's *Timaeus*, see Smith's "Introduction," in *Ptolemy's Theory of Visual Perception*, 22–23. Peter's inclusion of the cooperation of the viewer as one of the requirements in the process of vision is important because it emphasizes far more than what is seen in the work of other Perspectivists that the agency of the viewer is a necessity for completed vision. See below, chapt. 11.5, p. 134, for the optical discussion of the role of intention in the process of vision. The Latin word *species* is a technical term in Perspectivist optics, which I have retained to denote the visual image representing the form of the object that is viewed as it is conveyed through the visual medium and impinges on the eye.

It follows that two things are required for the forgiveness of someone who is sinful.*

The forgiveness of someone who is sinful requires the action of the free will and harmony with the grace of God.[2] For this reason, Augustine says: "He who created you without yourself will not forgive you without yourself."[3] As an illustration of this matter, the angel of Laodicea was spiritually blind, as is said in Revelation 3: "You say 'I am rich and in need of nothing,' and you do not know that you are wretched and miserable and poor and blind and naked."[4] The celestial physician said to the angel, "Anoint your eyes with a salve so that you can see."[5] By the salve, which cleanses and rinses the eyes, are signified contrition and remorse for sins and one's own efforts in carrying out good deeds: a human being cooperates with them in order to receive vision with his spiritual eyes. For this reason the *Gloss* says about the previously mentioned passage: "The eyes are anointed with a salve so that we may see, since in order to understand the clarity of the true light we sharpen our intellect with the medicine of good deeds."[6] Furthermore, as a sign of this matter, namely that we must cooperate in order to see spiritually, the Lord wanted Paul, who had been blinded, to pray and fast for three days before he regained his sight, as it is said in Acts 9.[7] The same thing is also indicated in the fact that "Tobias spread fish-gall on his father's eyes,"[8] "and immediately he regained his sight," Tobit 11.[9] In the fish-gall, which is bitter, are designated the bitterness and remorse caused by sins and the indignation against sin, for gall arouses anger. That is what is required on the part of someone who wants to be forgiven, so that his interior vision may be restored.

2 Cf. Roger Bacon, *Persp.*, 3.3.1, ed. and trans. Lindberg, 324. The close verbal parallels between Bacon's text and Peter's are noted in Newhauser, "*Inter scientiam et populum*," 701 (appendix, part D).

3 Augustine of Hippo, *Sermones*, 169.11.13, PL 38:923.

4 Rev 3:17.

5 Rev 3:18.

6 *Glossa ordinaria*, Apoc. 3:18, in *Biblia latina cum Glossa ordinaria. Facsimile Reprint of the Editio Princeps, Adolph Rusch of Strassburg 1480/81* (Turnhout, 1992), vol. 4, 554a in margin (PL 114:717C).

7 Cf. Acts 9:8–9.

8 Tob 11:13.

9 Tob 11:15.

CHAPTER FIVE

On the Organ in Which Vision Is Completed

As is taught in Perspectivist science, vision is not completed in the eyes, for two different *species* of the same object reach both eyes. Since a diversity in *species* creates a diversity in judgment, because of the perception of a double *species* one object would be judged to be two if vision were completed in the eyes. It is necessary, therefore, that another sentient organ be situated beyond the eyes in which vision can be completed, the instruments of which are the eyes that pass on to it the *species* of the visible object. This organ is the common nerve located on the surface of the brain, where the two nerves come together from the two parts of the anterior portion of the brain. After they come together, they are divided again into two, and in this form they are extended toward the eyes. In that common nerve, therefore, the visual power has its source originally. Since, then, this power is singular at its source, and the powers of the eyes form a continuum with it, for that reason one object can appear as a single thing, although it is seen by two eyes.[1]

That one should beware of rash judgment.*

In the fact that vision is not completed in the eyes, which appear externally, but rather in the common nerve, which is hidden internally, we are instructed morally to avoid rash judgment and not to judge matters as they appear at first sight, but rather to have recourse to an internal judgment through careful consideration. And for this reason concerning the angel of great counsel,[2] whose every action is our instruction according

1 Cf. Roger Bacon, *Persp.*, 1.5.2, ed. and trans. Lindberg, 62–64. The close verbal parallels between Bacon's text and Peter's are noted in Newhauser, "*Inter scientiam et populum*," 703 (appendix, part E).

2 I.e., Jesus; cf. Isidore of Seville, *Allegoriae quaedam Sacrae Scripturae*, 119, PL 83:115B. See also below, chapt. 15, p. 190.

to Gregory,[3] it is said in Isaiah 11: "Not according to the vision of his eyes will he judge."[4] For as Seneca says: "It is not enough to give one's attention to what is before the eyes."[5] For this reason many things often appear to be foul to the external eyes which nevertheless are clean if they are examined by the internal judgment of reason, a fact that is clearly evident concerning the judgment of leprosy, Leviticus 13: "But if spreading leprosy breaks out in the skin and covers all of the flesh from the head to the feet, and whatever can be seen with the eye, the priest will examine him and will judge that the leprosy he has is very clean*."[6]

3 Cf. Gregory the Great, *Dialogi*, 1.9.7, in *Dialogues*, trans. Paul Antin, ed. Adalbert de Vogüé, SC 260 (Paris, 1979), 80–81 (PL 77:192B). This statement is a commonplace for medieval theologians; cf. e.g. Peter Lombard, *In totum Psalterium commentarii*, *Psalmus* 29, 6, PL 191:295A.

4 Isa 11:3.

5 Cf. Boethius, *Philosophiae consolatio*, 2.pr.1.15, ed. Ludwig Bieler, CCSL 94 (Turnhout, 1957), 18.

6 Lev 13:12–13.

CHAPTER SIX

On 13 Extraordinary Phenomena That Are Related to Lessons Concerning Eyesight

It is taught in Perspectivist science, and experience teaches this as well, that an eye situated in foggy air can neither see nor perceive the mist and clouds that surround it; however, when it moves some distance away from this type of misty air and then looks back at it, it will see the misty air which it did not see earlier when it was located within it.[1]

In the same way, as long as a sinner is in a state of sin, he, too, does not notice the darkness of his sin, but once he is situated outside the sin and enlightened by the glow of divine grace, then for the first time he will recognize the magnitude of his sin and the mental blindness he suffered. This is exemplified by the ape lying next to a tree trunk who does not perceive the immensity of the trunk until he begins to move away from it.[2] And for the same reason, after David distanced himself through penance from the sin he had committed with Bathsheba, he understood his sin and said, "For I understand my iniquity."[3] As a typological symbol of this state it is said in Exodus 14: "Lifting up their eyes, the Israelites saw the Egyptians coming after them."[4] Now, Egypt is interpreted as "darkness,"[5] and it signifies sin, which is a spiritual darkness that a person understands for the first time when he retreats from sin by doing penance. For this reason, Anselm says in the *Similitudes*: "Vice is an immoral habit that is not easily seen by the person who is weighed down by it, but as soon as this person stops committing a vice and distances

1 Cf. Roger Bacon, *Persp.*, 1.9.1, ed. and trans. Lindberg, 126.

2 The text reads *ceperit elongari*, the passive of *elongare*, like the reflexive, meaning "to leave, to distance oneself." For apes exemplifying sinners, see Horst W. Janson, *Apes and Ape Lore in the Middle Ages and the Renaissance*, Studies of the Warburg Institute, 20 (London, 1952), 29–71.

3 Ps 50:5; cf. 2Sam 12:13.

4 Exod 14:10. For *figura* as a technical term in typological interpretation, see Erich Auerbach, "Figura," in *Scenes from the Drama of European Literature. Six Essays*, trans. Ralph Manheim (New York, 1959), 11–76.

5 Jerome, *Lib. interp. Hebr. nom.*, Act., ed. de Lagarde, CCSL 72:143. See below, chapt. 12.7, p. 168.

himself from it, then at last he carefully examines what misery and filth he was lying in."[6]

On the second extraordinary phenomenon concerning the eye's vision.

It is well known in the science mentioned already that if someone places a finger under his eye and pushes it from its normal position, because of the motion this creates in the eye a single object will appear to the viewer to be two objects.[7]

That evil advisors and bribes corrupt prelates*.

It sometimes happens in the same way to prelates of this modern age: When they are pushed from an upright judgment by the instigation of an evil advisor, they judge that someone is worthy of a double benefice, although if they were not deluded in their judgment they would feel that this person ought to be content with a single benefice alone. This often happens because of an evil advisor, who is properly represented as a finger because of the discernment with which he is believed to be endowed.[8] For with his perverse exhortations, as if with a kind of pressure, an evil advisor often pushes someone headlong into error "when he persuades him by twisted *auctoritates* and the examples of despots that all things are allowed to a magistrate or a prince and what is pleasing to a lord has the force of law. And as examples he mentions which despot forced his servant or a member of his household onto what church without an election, which one levied taxes in the dioceses, which one ruined the clergy, which one forced the bishops into silence, finally, which one satisfied the pleasure of his own will on his subjects and made

6 Ps.-Anselm of Canterbury, *Liber de similitudinibus*, 99, PL 159:665A (cf. Alexander of Canterbury, *Liber ex dictis beati Anselmi*, 1, ed. Richard W. Southern and Franciscus S. Schmitt, in *Memorials of St. Anselm*, Auctores Britannici Medii Aevi, 1 [Oxford, 1969], 110).

7 Cf. Roger Bacon, *Persp.*, 1.5.2, ed. and trans. Lindberg, 64. Bacon uses the experiment of pushing against the eye to create a dual image to argue that the visual *species* from the eyes must converge at a single point in the common nerve to achieve a singular image. Moving the eye from its natural place will make the *species* from that eye arrive at a different location in the common nerve than the *species* from the other eye. Peter, on the other hand, attributes the visual effect to the motion created in the eye itself.

8 Cf. Gregory the Great, *Mor. in Iob*, 15.37.43, ed. Adriaen, CCSL 143A:776.

what he himself desired the same as what is lawful."[9] Hence, the verse reads:

> Thus I desire, thus I command; let my will be reason enough.[10]

And just as an evil advisor corrupts kings and princes, so too he sometimes corrupts even bishops and other leading figures of the church. For if a prelate has been asked about the advancement of someone who is unworthy and objects that this person has some weakness and therefore cannot rightly be promoted, the evil advisor in attendance will demolish whatever objection has been made, "with *exempla* and the decrees of the fathers that are at variance with it.[11] For if the prelate were to object, 'The person is of low birth,' the evil advisor will respond, 'But Peter, the pinnacle of the church, never bragged about the splendor of his blood line.'[12] If it is said that the person is too young, he will respond that Jeremiah and the Lord's forerunner were boys, and, again, he will mention the elders damned by Daniel, a boy.[13] 'The person is illiterate, and he's never been to Paris or any other city to study' – and he will respond, 'Nor does one read that the apostles frequented schools.' 'He's lacking in eloquence' – 'But as one reads in Exodus, Aaron, too, performed the priestly office in place of Moses, who (that is to say, Moses) was of very halting speech.'[14] 'He's lustful and frequents whorehouses' – 'But Hosea copulated with a whore when the Lord ordered him to do so.'[15] 'He's a fool' – 'But God decreed that believers be saved through the foolishness of the world.'[16] 'He's a thug and has carried weapons' – 'But Peter, too, "cut off the ear of the chief priest's servant with a sword."'[17] 'He's a drunkard and given to gluttony' – 'But one reads that the Lord

9 John of Salisbury, *Policraticus*, 7.20, ed. Clemens C.I. Webb, in *Policratici sive De nugis curialium et vestigiis philosophorum libri viii* (Oxford, 1909; reprint New York, 1979), vol. 2, 187–88.

10 Juvenal, *Saturae*, 6.223, ed. Wendell V. Clausen, rev. ed. (Oxford, 1992), 79 (cf. Walther, *Proverbia*, no. 29559).

11 For the source of the long passage beginning here and ending with "as God also did at that time," see below at n35.

12 Cf. 2Pet 1:1.

13 Cf. Jer 1:6–7, cf. Luke 1:17 (on John the Baptist), 1:66, 1:80; cf. Dan 13:45–49.

14 Cf. Exod 4:10–14.

15 Cf. Hos 1:2–3.

16 Cf. 1Cor 1:21, 27.

17 Matt 26:51, cf. John 18:10.

himself said that he was a drinker of wine and an eater of meat,[18] "and no servant is greater than his lord."'[19] 'He doesn't submit to his superiors, nor is he willing to obey them in anything' – 'But one reads that Paul, too, opposed Peter, the chief of the apostles, to his face.'[20] 'He's contentious and a disturber of the peace' – 'But among the disciples of Jesus Christ "contention arose about which one of them was to be regarded as the greater."'[21] 'He's carried arms in warfare' – 'But Martin, too, served for a time as a soldier under Julian.'[22] 'He's a murderer and has spilled blood' – 'But Moses, too, killed an Egyptian.'[23] 'He's dishonest and has broken his word frequently' – 'But the apostle Peter also combined betrayal with perjury.'[24] 'He's mute and would be unable to preach' – 'But that didn't exclude Zechariah from the priesthood.'[25] 'He's blind' – 'But Paul, too, had been blinded when he was baptized by Ananias and consecrated at the Lord's command.'[26] 'He's deaf' – 'But that doesn't prevent him from preaching God's law. For we seek a preacher, not a listener, and he may proclaim the word of God all the more loudly, and he'll raise his voice like a trumpet[27] in his sermons, since it'll still seem to him that he's only speaking softly into someone's ear.' 'His character is despicable' – 'But Defensor likewise objected to Saint Martin for this reason; nevertheless, the church didn't despise him.'[28] 'He's sickly' – 'But although Gregory had the most debilitating stomach ailment, he governed the church of God in the best way.'[29] 'He's proud and vain' – 'But so was Brice*, archbishop of

18 Cf. Matt 11:19.

19 John 15:20.

20 Cf. Gal 2:11.

21 Luke 22:24.

22 Cf. Sulpicius Severus, *Vita s. Martini*, 2.2, ed. Karl Halm, in *Libri qui supersunt*, CSEL 1 (Vienna, 1866), 111. The reference is to Saint Martin of Tours (ca. 316–ca. 397) who was enrolled in the Roman army as a youth and served under Emperor Julian ("the Apostate"; 331–363).

23 Cf. Exod 2:12.

24 Cf. Matt 26:69–75.

25 Cf. Luke 1:22.

26 Cf. Acts 9:8–17.

27 Cf. Isa 58:1.

28 Cf. Sulpicius Severus, *Vita s. Martini*, 9, ed. Halm, CSEL 1:119, who relates that Defensor was one of the bishops who attempted to prohibit Saint Martin from becoming the bishop of Tours.

29 Cf. Paul the Deacon, *Vita beatissimi Gregorii papae urbis Romae*, 13, ed. Hartmann Grisar, "Die Gregorbiographie des Paulus Diakonus in ihrer ursprünglichen Gestalt, nach italienischen Handschriften," *Zeitschrift für katholische Theologie* 11 (1887): 158–72, here 170. The reference is to Pope Gregory the Great (ca. 540–604).

Tours.'[30] 'He's sometimes taught heresy' – 'But Augustine declared that he had been a Manichee.'[31] 'He's persecuted the church of God' – 'But Paul, too, earlier a persecutor, was later made a preacher.'[32] 'He's avaricious' – 'In this way he'll be able to gather up what has been scattered and he won't uselessly squander what has been gathered.' Finally, if the prelate objects that in all these matters the person is clearly unsuitable, the evil advisor will respond to this, 'But Samson, too, swept away the Philistines with the jawbone of an ass.[33] And now, too, God has the power "to raise up children for Abraham from stones,"[34] as God also did at that time.'"[35]

You see, thus, how powerful this finger is, that is to say, an evil advisor, because by the pressure of his persuasion he can make the eye of the observer, i.e., a prelate, go astray in seeing and in corrupting his judgment. And in this, indeed, the more discerning he is thought to be, the more powerful will he be regarded, so that it may be said about him, "He is the finger of God."[36] Ahithophel, Absalom's most corrupt advisor, had this characteristic, and nevertheless it is said about him in 2 Samuel 16: "The advice of Ahithophel, which he gave in those days, was as if someone were to consult God."[37] For what is this if not to say, "He is the finger of God"? For that reason it is said in Ecclesiasticus 37: "From an evil advisor preserve your soul."[38] And David prayed, "Lord, deliver my soul from wicked lips and from a deceitful tongue."[39] These words, too, a prelate should offer in prayer, and request from the Lord.

Now, the eye of the church, i.e. a prelate, is deceived not only by an evil advisor in the manner just mentioned, but also sometimes by the acceptance of bribes, which in a certain manner knock the eye of reason from its place. And so in common parlance we are accustomed to say that whoever has already accepted bribes is crooked and, hence, that he perverts his judgment. These are the fingers about which it is said in Isaiah

30 Cf. Gregory of Tours, *Libri historiarum decem*, 2.1, ed. Bruno Krusch and Wilhelm Levison, MGH, Scriptores rerum Merovingicarum, 1/1, 2nd ed. (Hannover, 1951), 37. Saint Brice (ca. 370–444) succeeded Martin as the bishop of Tours in 397.

31 Augustine of Hippo, *Confessiones*, 3.6.10, ed. Luc Verheijen, CCSL 27 (Turnhout, 1981), 31; idem, *De utilitate credendi ad Honoratum liber unus*, 3.7, PL 42:69.

32 Cf. 1Tim 1:13, 2:7.

33 Cf. Judg 15:15.

34 Cf. Luke 3:8.

35 The entire passage reporting the conversation between the prelate and the evil advisor is taken from John of Salisbury, *Policr.*, 7.19, ed. Webb, vol. 2, 175–78.

36 Exod 8:19.

37 2Sam 16:23.

38 Sir 37:9.

39 Ps 119:2.

59: "Your fingers are defiled by iniquity,"[40] and hence it is no wonder if they make the eye of reason deviate from a balanced judgment. For, indeed, above all three things normally corrupt the judgment of someone who is wise, namely an inclination to believe others, the love of bribes, and showing partiality.[41] Thus, it is said in Deuteronomy 16: "Gifts blind the eyes of the wise and alter the words of the just."[42] And about the sons of Samuel* it is said in 1 Samuel 8: "They accepted bribes and they perverted justice."[43] And the poet says:

> For a gift distracts a judge's sagacity from the
> Standards of law, and envelops his mind in a shameful fog.[44]

Therefore, one must mourn and lament if it happens that prelates, who are the judges of their age and the true lights of the world, love bribes. May God prevent it from happening! May retributions follow if they pillage their subjects in order to snatch their purses, if they empty others' purses in order to fill up their own, so that it can be said, "In their hands are evils, whose right hand is filled with bribes!"[45]

Therefore, let the prelates of the church shake "their hands from every bribe"[46] if they desire to be unerring in their judgments, following the example of Eugene who "accepted absolutely no bribe from a person bringing litigation nor from one against whom he believed litigation was pending. For this reason he is also reported to have said, and not without great indignation, to a certain prior who was moderately rich, whose case he had not yet heard, and who was most reverently offering him a mark of gold, 'You have not yet entered the house, and you already want to

40 Isa 59:3.

41 Cf. 2Chr 19:7.

42 Deut 16:19.

43 1Sam 8:3.

44 Walter of Châtillon, *Alexandreis*, 1.109–10, ed. Marvin L. Colker, Thesaurus mundi, 17 (Padua, 1978), 13 (cf. Walther, *Proverbia*, no. 15729b).

45 Ps 25:10. The passage from "judges of their age" to the end of the paragraph is adopted from John of Salisbury, *Policr.*, 8.17, ed. Webb, vol. 2, 355. The foundation of the wordplay in describing prelates as pillaging (*concuciant*) those entrusted to them in order to snatch (*excuciant*) their purses is found in the passage from Isaiah quoted at the beginning of the next paragraph, in which the man who lives an upright life is said in the Vulgate text to shake (*excutit*) his hands from bribes. This kind of wordplay based on near homonyms was clearly attractive to Peter of Limoges and is found elsewhere in *The Moral Treatise on the Eye*. See, for example, below, chapt. 6.3, p. 27.

46 Isa 33:15.

corrupt the house?' In *On Consideration* Bernard also relates to Eugene something concerning the cardinal-priest Martin*: When he was returning from a legation as a pauper, contrary to the custom of the times, Martin accepted a horse from the bishop of Florence which was indispensable for him, but he returned the horse to its donor since he did not know that at the moment the bishop had given him the gift, he had a court case that was being heard in the church at Rome. 'You have deceived me,' Martin said; 'I did not know that you have a case pending. Take back your horse immediately.'[47] Behold the statement of a man who hated the acceptance of bribes! Geoffrey of Chartres, too, accepted no gifts while he carried out the office of legate in Aquitaine, but rather hated them like excrement. He agreed at long last to accept a sturgeon, namely from one of the devout clerics in his legation, only after he counted out the price for the gift he had allowed."[48]

If modern prelates would act in this way, they would be loved like fathers, feared like lords, and cherished like saints: if they collected taxes sparingly, did not love bribes, and did not consider every acquisition a matter of piety. But today "in the houses of many prelates and princes all things are for sale, since avarice rules there.[49] For:

> If once the mother of vices gains a hold on the court –
> The plague of avarice, which by itself imprisons every kind of
> virtue –,
> Then while the rule of law is scorned everyone will run*
> To commit crimes, nor will the court preserve the laws.[50]

Only in vain will you trust in the testimony of your conscience or in your pleasing manners among courtiers of this kind if you have not prepared the way by sending a bribe ahead of you. As Ovid says:

47 Cf. Bernard of Clairvaux, *Tractatus de consideratione ad Eugenium papam*, 4.5.13, ed. Jean Leclercq and Henri M. Rochais, in *Opera* (Rome, 1957), vol. 3, 459. Martin, a cardinal-priest of the Basilica of Saint Stephen on the Celian Hill in Rome near the middle of the twelfth century, had earlier been a monk at Clairvaux.

48 Cf. Bernard of Clairvaux, *Tract. de consid.*, 4.5.14, ed. Leclercq and Rochais, in *Opera*, vol. 3, 459. The entire passage from "accepted absolutely no bribe" to the end of the paragraph is taken from John of Salisbury, *Policr.*, 5.15, ed. Webb, vol. 1, 347–49. The Eugene referred to here is Pope Eugene III (d. 1153). Geoffrey of Lèves (d. 1149), an important reformer in the French church and an acquaintance of Bernard of Clairvaux, succeeded Ivo as bishop of Chartres in 1116.

49 For the passage beginning "in the houses of many prelates" and ending with "either barking or biting," see below at n52.

50 Walter of Châtillon, *Alex.*, 1.111–14, ed. Colker, 13 (cf. Walther, *Proverbia*, no. 4411).

Should you yourself, Homer, arrive accompanied by the Muses,
If you haven't brought anything with you, you will go right back out
the door again, Homer.[51]

In many courts you will find gatekeepers more unfeeling than three-headed Cerberus, but among the inhabitants of hell there is said to be only one Cerberus. Here, on the other hand, there are as many Cerberuses as there are door-keepers; you will find a whole household of them either barking or biting."[52] And as it is said in Micah* 3: "If a person does not put something in their mouths, they prepare war against him."[53] But:

Once the coin has been heard from, as if the greatest prince had
just come into view,
The doors will fly open; nothing is heard except: "Welcome!"[54]

One reads in the life of the blessed Anselm that he used to shake "his hands from every bribe,"[55] fleeing from those who accepted bribes as from people who are detestable. And he also said to them frequently, "'To take (*prendere*) and to hang (*pendere*) do not differ from each other except in one letter alone.' For this reason it is obvious that whoever accepts gifts freely is just a step away from being hanged, unless he accepts them in the right way."[56] Above all, however, each person should beware of receiving

51 Ovid, *Ars amatoria*, 2.279–80, ed. Edwin J. Kenney, in *Amores, Medicamina faciei femineae, Ars amatoria, Remedia amoris*, 2nd ed. (Oxford, 1994), 164 (cf. Walther, *Proverbia*, no. 12837).

52 For the passage beginning "in the houses of many prelates," cf. Vincent of Beauvais, *Spec. doct.*, 7.22, in *Speculum quadruplex*, vol. 2, 572a–b. Cerberus, the three-headed dog who guarded hell in Greek and Roman mythology, was said to have allowed none of the souls to leave once they entered hell.

53 Mic 3:5.

54 *Sententia communis* (Walther, *Proverbia*, no. 1740).

55 Isa 33:15.

56 *Vita beati Edmundi Cantuariensis Archiepiscopi et confessoris*, 53, ed. Edmond Martène and Ursin Durand, in *Thesaurus novus anecdotorum* (Paris, 1717), vol. 3, 1807B–C. Peter has Latinized the near homonyms used by Saint Edmund (not Anselm, to whom the well-known anecdote is mistakenly attributed in both *P7P8* and *Pr1*). The French forms are often reported by other authors, as they are in the life edited by Martène and Durand, with which Peter's text otherwise agrees verbatim. See also the excerpts of the Balliol life of Edmund in Henry W.C. Davis, "An Unpublished Life of Edmund Rich," *The English Historical Review* 22, no. 85 (1907): 84–92, here 89: "Entre prendre et pendre ni ad ke une lettre. Dunt cil ke prent volunters est mut pres de pendre." Cf. also Thomas de Eccleston, *Tractatus de adventu fratrum minorum in Angliam*, collatio 15, ed. Andrew G. Little (Man-

gifts from those who are evil, lest, having accepted the gift, he be censured as an ingrate if he does not favor them, or be thought of as unjust if he does favor them.

On the third extraordinary phenomenon. On the examination of sin and its double punishment.*

Among the Perspectivists it is well known that if at night someone raises a branch or a finger (or any other kind of object that casts a shadow) between his eyes and a candle and then looks directly at the candle, it will appear to him that where there was one finger, there are now two. And along with this, one ought to observe attentively that if the right eye is closed, the image on the left will disappear, and with the left eye closed, the image on the right vanishes completely. Augustine marveled quite a bit about this matter, and for this reason he says in book 11, chapter 2 of *On the Trinity*: It takes a long time to explain the cause of this optical phenomenon,[57] and it takes a truly long time for a person with no knowledge of Perspectivist science.[58]

In a similar fashion, if a sinner living in the night of sin places his sin before the eye of his contemplation and looks at the lamp of the word of God, concerning which the Psalm says: "Your word is the lamp for my feet,"[59] he will see that each of his sins is in a sense doubled if one takes into account the punishment that corresponds to it. He will see (as long as he is able to be attentive to the light of the word of God) that he has incurred a double punishment for a single sin, in such a way that the guilt requiring torment, so to speak the image on the left, appears to the right eye, with which he should have performed good deeds, while the trophy of heaven that he has lost, like the image on the right, appears to the left eye, with which evil should have been avoided.

The sinful soul is warned to avoid this appalling and extraordinary sight in Jeremiah 2: "Know and see how evil and bitter it is for you to

---

chester, 1951), 93: "Item dixit idem sanctus episcopus, cum offerrentur sibi quaedam pretiosa iocalia, et moneretur a suis ut ea reciperet, respondit: Si prenderem, penderem; entre prendre e pendre non est nisi una litera."

57 Cf. Augustine of Hippo, *De trinitate*, 11.2.4, ed. William J. Mountain and François Glorie, CCSL 50 (Turnhout, 1968), 337–38.

58 Cf. Roger Bacon, *Persp.*, 2.2.3, ed. and trans. Lindberg, 188. The close verbal parallels between Bacon's text and Peter's are noted in Newhauser, "*Inter scientiam et populum*," 703 (appendix, part F).

59 Ps 118:105.

have left the Lord your God."[60] To have left the Lord is nothing other than to have sinned, and this sin, although perpetrated (*committatur*) by a single act, is nevertheless accompanied (*comitatur*) by a twin punishment. For it is evil in terms of the deprivation of divine vision and bitter because of the harshness of punishment in hell. And since this extraordinary phenomenon is plainly apparent if someone contemplates the lamp of the word of God, therefore the next passage in the same chapter of Jeremiah says expressly: "See the word of the Lord,"[61] namely so that you will have the power to observe the extraordinary sight* just mentioned.

The fourth extraordinary phenomenon. That prelates should beware of becoming a bad example or an enticement to sin.

It is well known in the previously mentioned science that when one part of a rod is in water while the rest sticks out above the water, the rod will appear broken when the eye is located in the air. The cause of this phenomenon, as is demonstrated in this same science, is given as follows: An object that we look at in water appears to the eye to be closer than it is, measured by the actual distance of its location, and thus the part of the rod in the water does not appear to our sight as a continuous and straight extension of the other part, but seems closer to the eye itself, and so the rod appears to be broken.[62]

60 Jer 2:19. Note the wordplay in Peter's description of the soul's double punishment for a single sin as being committed (*committatur*) by a single act, but accompanied (*comitatur*) by a double punishment.

61 Jer 2:31.

62 Cf. Roger Bacon, *Persp.*, 3.2.4, ed. and trans. Lindberg, 308–10; cf. John Pecham, *Perspectiva communis*, 3.6, ed. and trans. David C. Lindberg, in *John Pecham and the Science of Optics: Perspectiva communis* (Madison, etc., 1970), 216 (see Gudrun Schleusener-Eichholtz, "Naturwissenschaft und Allegorese: Der 'Tractatus de oculo morali' des Petrus von Limoges," *Frühmittelalterliche Studien* 12 [1978]: 258–309, here 290n160). Peter's explanation of refraction is taken partially verbatim from both works. Compare, e.g., the following phrase in Peter's text (in the wording of *P7P8*): "et ideo pars baculi existens in aqua non apparet uisui in continuum et directum partis alterius, sed ipsi oculo apparet esse propinquius" (and thus the part of the rod in the water does not appear to the sight as a direct extension of the other part, but appears to be closer to the eye itself) with its equivalent in Bacon's text: "Et ideo pars baculi que est in aqua non apparebit visui in continuum et directum alterius partis, sed propinquius oculo"; and Peter's treatise (following *P7P8*): "res, quam in aqua conspicimus, propinquior apparet oculo quam sit secundum ueram distanciam sui situs" (an object which we look at in water appears to the eye to be closer than it is, measured by the actual distance of its location) with its equivalent in Pecham's work: "pars existens in aqua propinquior apparet quam sit secundum veritatem."

It happens occasionally in the same way if someone who is truly an upright and God-fearing person should ever, and for whatever reason, make use of the pleasures of this world, which are close to the eyes of worldly people. Although he does not immerse himself completely in the stream of these kinds of pleasures, but only up to a point, since presumably he does not act this way because of the wantonness of his mind, but because of the enjoyment his body finds necessary, nonetheless he sometimes leads many others who see this into temptation, and he is judged by common opinion to be broken spiritually and to have turned aside from moral rectitude.

And using this phenomenon we can explain what one reads in 1 Samuel 18 that Saul said concerning David, "'I will give him Michal,' so that she might become an enticement for him to sin."[63] Michal is interpreted as "every water,"[64] and she signifies the stream of carnal pleasures. If any perfect man (symbolized by David) unites himself with these pleasures in love, he is turned into an enticement for many others to sin. And thus those who are perfect ought to be diligently on guard against ostentatious and dissolute acts of this kind in order to avoid being an enticement for others to sin. The apostle Paul, who urged others to imitate him, was paying attention to this fact when he said in 1 Corinthians 8, "Yet be careful, lest perhaps this liberty," namely of eating meat consecrated to idols, "become a pitfall for the weak."[65] And Paul adds, "For that reason, if food is a temptation for my brother to sin, I will never eat meat, lest I become an enticement for my brother to sin."[66] For the Apostle knew that the deeds of lords are easily taken as examples by their subjects: The more elevated the place which those of higher rank hold among the people, the greater is the number of those* to whose sight they are exposed. And for that reason the common man tries eagerly to be like his superior. And so Claudian says:

> According to the example of its king does the whole world shape itself.
> The unstable mob always changes with its prince.[67]

63 1Sam 18:21.

64 Jerome, *Lib. interp. Hebr. nom.*, 1 Reg., ed. de Lagarde, CCSL 72:104.

65 1Cor 8:9.

66 1Cor 8:13.

67 Claudius Claudianus, *Paneg. dict. Hon. Aug. quart. cons.*, 299–300, 302, ed. Hall, 72 (cf. Walther, *Proverbia*, nos. 14986 [the second verse] and 26481 [the first verse]).

The fifth extraordinary phenomenon. From which it follows that sins of the flesh are of greater disrepute than spiritual sins.

It is proven in Perspectivist science that things seen in water appear larger to the eye of the observer, and occasionally some things that the eye could not see at first can be seen when water is poured over them. And anyone can experience this with his senses. For if a penny is placed in a small, shallow dish and moved* to a distance from which it can barely be seen, if the distance between the eye of the observer and the shallow dish remains unchanged, it will be possible to see the penny when water is poured over it.[68]

In a similar fashion we can see that carnal sins, which are committed in the stream of pleasures, seem larger to the eyes of human beings, for these sins are of greater disrepute than spiritual sins although less blame attaches to them and it is hardly possible to keep them hidden. And frequently, when sinners ensnared by carnal sins think their sin is not seen by others, it comes out in public and is known to all. And for this reason when David believed that the sin he had committed with Bathsheba was hidden from everyone, the prophet Nathan was sent to him in order to say on behalf of the Lord, "You acted in secret, but I shall speak this word openly in the sight of all Israel and in broad daylight," 2 Samuel 12.[69] And in Isaiah 47 this is what the Lord threatens such people with: "Your disgrace will be revealed and your dishonor will be seen,"[70] and in Nahum 3: "I will reveal your private parts to your embarrassment, and I will expose your nakedness to the nations and your disgrace to every kingdom."[71]

The sixth extraordinary phenomenon. From which it is educed that Holy Scriptures are seen directly by the person who fulfills them in his deeds.

It has also been demonstrated in the previously mentioned science that the eye perceives an object that it sees in a mirror with weaker vision than if it were to discern it by looking at it directly, because reflected forms are

68 Cf. Roger Bacon, *Persp.*, 3.2.4, ed. and trans. Lindberg, 312, quoting Euclid. See Euclid, *De speculis*, assumption 6, ed. and trans. Ken'ichi Takahashi, in *The Medieval Latin Traditions of Euclid's Catoptrica. A Critical Edition of De speculis with an Introduction, English Translation and Commentary* (Fukuoka-shi, 1992), 114–15.

69 2Sam 12:11–12.

70 Isa 47:3.

71 Nah 3:5.

weaker and thus they have a weaker image. And, hence, a person scarcely and poorly recollects his own form that he often sees in a mirror, but he can picture to himself another person's face that he has looked at directly a few times better than his own that he has seen in a mirror, but never directly.[72]

And likewise, in its moral interpretation, by "the face" I understand divine law. For just as a person's internal will is signaled in his face according to the verse:

> In a person's face you can read his hidden will,[73]

so also God's will is manifested to us in divine law. And just as the face gives evidence of the person, according to the passage in Ecclesiasticus 19: "You can tell a person by his appearance, and a wise man is known by the first sight of his face,"[74] so, too, do Holy Scriptures give evidence of Christ. For this reason in John 5 the Lord said, "Search the Scriptures: They are the texts that produce testimony of me."[75] This divine face, i.e., Holy Scriptures, is seen directly by the person who fulfills Scriptures in his deeds and then retains them well in memory. But he who only hears Scriptures and does not carry them out to their conclusion perceives them as if in a mirror, and this person then consigns them immediately to oblivion. And therefore it is said in James 1: "Be those who carry out the word and not merely those who listen to it, deceiving yourselves, since if someone listens to the word and does not carry it out, he is comparable to a person contemplating in a mirror the face he has had from birth. For he has examined himself and has gone on his way and has immediately forgotten what he looked like."[76] And therefore if someone wants to make progress in Holy Scriptures and to arrive at a state of being knowledgeable, let him be eager to fulfill in his deeds the things he has read in Scriptures. And therefore David shows by which path he achieved wisdom, saying, "I understand more than those who are old, since I have sought to learn your commandments."[77] And again: "From your commandments I have acquired understanding."[78] The *Gloss* says: "Through

72 Cf. Alhacen, *De aspect.*, 4.2 [2.9–10], ed. and trans. Smith, in *Alhacen on the Principles of Reflection*, vol. 1, 6 (Latin text), vol. 2, 297 (English translation).

73 *Sententia communis* (Walther, *Proverbia*, no. 11765).

74 Sir 19:26.

75 John 5:39.

76 Jas 1:22–24.

77 Ps 118:100.

78 Ps 118:104.

obedience to the commandments he has achieved wisdom in hidden matters."[79]

Therefore, a person ought to have practice in the effort needed for good deeds before he takes up the effort to contemplate truth, for these two efforts are connected. Nor should he separate one from the other, so that, for instance, he might expend his effort in achieving wisdom, but not a good life. The effort needed for good deeds, however, should precede the effort needed for truth, and for this reason the Psalm says: "Teach me goodness and discipline and knowledge, for I have believed your commandments."[80] And hence the verse "Happy are they whose way of life is pure, who walk in the law of the Lord"[81] explicitly precedes the verse that immediately follows it: "Happy are they who investigate his testimonies."[82] Thus it is said in Ecclesiasticus 1: "My son, if you desire wisdom, preserve justice, and God will grant it to you."[83]

On the seventh extraordinary phenomenon. On the truth of a trinity and unity.*

Among the Perspectivists it has been proven and it is commonly known that if the eye looks at the sun in a mirror placed in water, there appear to be two suns; indeed, in this optical phenomenon* there are, as it were, three suns which seem to come together, namely one existing in the sky and two proceeding from it, i.e., two images of the sun appearing on the water and in the mirror. And nevertheless we know that according to the object itself and the truth of the matter there is only one sun.[84]

In similar fashion, if we use the eye of our faith to contemplate the sun of justice in the water of the wisdom that brings salvation, God will appear to us as threefold and singular. For this reason at the very commencement of divine wisdom God is represented as saying, "Let us make human beings in our image and likeness."[85] "By saying 'let us make' and

79 *Gloss. ord.*, Ps. 118:104, ed. Rusch, vol. 2, 615b in margin. Cf. Augustine of Hippo, *Enarrationes in Psalmos, In Ps. 118, sermo* 22, 8, ed. Eloi Dekkers and Jean Fraipont, CCSL 40 (Turnhout, 1956), 1740.

80 Ps 118:66.

81 Ps 118:1.

82 Ps 118:2.

83 Sir 1:33.

84 Cf. Roger Bacon, *Persp.*, 3.1.6, ed. and trans. Lindberg, 284. The phenomenon that Peter of Limoges and Roger Bacon refer to here is the reflection of one image of the sun in the surface of the water and another image of the sun in the mirror submerged in the water.

85 Gen 1:26.

'our,' he showed a plurality of persons, but by saying 'in the image and likeness' he demonstrated the unity of his essence."[86] And innumerable testimonies declaring the truth of an ineffable unity and trinity can be made clear to anyone examining the sacred text.

In the similitude just mentioned I judge this worthy of consideration: Since a double image is reflected in the mirror, one reads that in dealing with one of these images quite a few Perspectivists have erred by considering one of them to be an image of some star located, as they believe, near the sun. This is something, however, which is false and erroneous, for it is not a star which appears, but rather a double image of the sun reflected by two surfaces, as must be demonstrated elsewhere.[87]

In similar fashion, in dealing with the third person of the Trinity, i.e., the Holy Spirit, one reads that quite a few have erred. As an indication of this matter, in Exodus 8 the magicians are said to have failed in [producing] a third sign.[88]

The eighth extraordinary phenomenon. That God is a mirror without blemish.*

It has been proven in Perspectivist science that if "the eye is in the center of a concave spherical mirror and it looks at the mirror, it sees only itself." For this reason if the entire heavens were a mirror and if an eye in the center of it looked at the heavens, it would see itself alone. This matter is proven as follows: "Since an object located outside the center has rays which fall obliquely onto the surface of the mirror, and since the angles of reflection and incidence must be equal, it follows that the rays are reflected not to the center itself but to the opposite side."[89]

86 Peter Lombard, *Sententiae in iv libris distinctae*, 1.2.4.2, ed. PP. Collegii s. Bonaventurae, 3rd ed., Spicilegium Bonaventurianum, 4 (Grottaferrata, 1971), 64.

87 Cf. Roger Bacon, *Persp.*, 3.1.6, ed. and trans. Lindberg, 284. Here, again, Peter's text is taken partially verbatim from Bacon's: compare, e.g., "Non est enim stella que apparet, sed est duplex solis ymago a duplici reflexa speculo" (for it is not a star which appears, but rather a double image of the sun reflected by a double mirror) in Peter's text, following the wording of *P7P8*, to the equivalent in Bacon's work: "Quapropter non est stella que apparet, sed est duplex ymago solis vel lune vel candele de duplici speculo reflexa." The "double mirror" referred to here, as Bacon demonstrates at greater length, is the combination of the surface of the water and the mirror submerged in the water, each of which offers a reflection of the sun. *Pr1* renders this as two reflective surfaces.

88 Cf. Exod 8:18. Cf. Peter Lombard, *Sent.*, 1.3.1.9, vol. 1, 71.

89 John Pecham, *Persp. comm.*, 2.41, ed. and trans. Lindberg, 194 (see Schleusener-

So, too, is it for God to see, in his own way, for as Alan says: "God is an intelligible sphere whose center is everywhere."[90] He is also a "mirror without blemish," as it is said in Wisdom 7.[91] Since the divine eye is the innermost part of each object – being as it were, the thing most present in the center of all things –, therefore, as if in a mirror, it alone understands itself within itself and through itself. But one does not say that God has a knowledge of things within themselves other than himself, lest, as Aristotle says, his intellect become tainted,[92] but he knows and understands these other objects within himself, insofar as he is an exemplar of all things, containing in his essence an ideal likeness of each individual thing.

On the ninth extraordinary phenomenon, from which it follows that encumbered by our body, we cannot see the divine essence.*

It is also proven in the previously mentioned science that when an object is in the center of a concave spherical mirror, it is not seen by an eye located outside its center when it looks at the mirror.[93]

Eichholtz, "Naturwissenschaft," 288n145). Cf. Roger Bacon, *Persp.*, 3.1.4, ed. and trans. Lindberg, 272. The equivalence of the angles of incidence and reflection is discussed, e.g., by Bacon in *Persp.*, 3.1.1, ed. and trans. Lindberg, 254, with explicit mention of Euclid's *Elements*. For Euclid's statement on the eye in the center of a concave mirror, see *De spec.*, proposition xxiv, ed. and trans. Takahashi, 172–75. Cf. on concave spherical mirrors, Alhacen, *De aspect.*, 4.5 [5.46–50], ed. and trans. Smith, in *Alhacen on the Principles of Reflection*, vol. 1, 59–60 (Latin text), vol. 2, 342–43 (English translation); and on the equivalent angles of incidence and reflection, *De aspect.*, 4.3 [3.2], vol. 1, 9 (Latin text), vol. 2, 300 (English translation).

90 Alan of Lille, *Sermo de sphaera intelligibili*, ed. Marie-Thérèse d'Alverny, in *Textes inédits*, Études de philosophie médiévale, 52 (Paris, 1965), 295–306, here 297; idem, *Theologicae regulae*, 7, PL 210:627A.

91 Wis 7:26.

92 For the idea that God knows other things only in himself lest his intellect become tainted (*eius vilesceret intellectus*), with references to Aristotle's *Metaphysics*, cf. Alexander of Hales (and his students), *Summa theologica. Pars prima*, inq. 1, tract. 5, sect. 1, quaest. unica, memb. 3, cap. 6, 2, in *Summa theologica*, ed. PP. Collegii s. Bonaventurae (Quaracchi, 1924), vol. 1, 256; Albert the Great, *Summa theologiae. Pars prima*, tract. 15, quaest. 60, memb. 3, ad quaest. 4, ed. Auguste Borgnet, in *Opera omnia* (Paris, 1894), vol. 31, 609. I am grateful to Bonnie Kent for bringing these passages to my attention. Cf. the discussion in Aristotle, *Metaphysica*, 12.9, trans. William of Moerbeke, ed. Gudrun Vuillemin-Diem, Aristoteles Latinus, 25.3.2, Corpus Philosophorum Medii Aevi, 1.25.3.2 (Leiden, etc., 1995), 264–66; and ibid., translatio anonyma sive "Media," ed. Gudrun Vuillemin-Diem, Aristoteles Latinus, 25.2, Corpus Philosophorum Medii Aevi, 1.25.2 (Leiden, 1976), 219–21.

93 Cf. Roger Bacon, *Persp.*, 3.1.4, ed. and trans. Lindberg, 272; cf. John Pecham, *Persp.*

Thus we, too, are not capable of seeing the divine essence as long as our sojourn in this life is prolonged[94] and we are separated from God by the heavy weight of our oppressive flesh. And for this reason the Lord answered Moses, who prayed that he might see the face of God, "You cannot see my face, for no human being will see me and live," Exodus 33.[95] When, however, we hasten to achieve understanding in order to fulfill what one reads in 1 Corinthians 6: "Whoever joins himself to God is one with him spiritually"[96] – at which time, too, "God is all in all"[97] –, then no longer outside the center but, as it were, placed in the center of divine steadfastness and, in a certain sense, made one with God through love, "we will see him just as he is," 1 John 3.[98]

On the tenth extraordinary phenomenon, from which it is proven that what the world considers favorable is unfavorable in God's sight, and vice versa.*

To the eye of someone looking in a mirror, shapes appear in reverse and vertical dimensions seem to be upside down. For when it perceives objects in a mirror, the eye judges that what is right is left and vice versa, and what is above is below and what is below is above.[99]

---

comm., 2.40, ed. and trans. Lindberg, 192 (see Schleusener-Eichholtz, "Naturwissenschaft," 288n147). Cf. Alhacen, *De aspect.*, 4.5 [5.48], ed. and trans. Smith, in *Alhacen on the Principles of Reflection*, vol. 1, 59–60 (Latin text), vol. 2, 342 (English translation); Euclid, *De spec.*, proposition xxvi, ed. and trans. Takahashi, 176–79.

94 Cf. Ps 119:5.

95 Exod 33:20.

96 1Cor 6:17.

97 Cf. 1Cor 15:28.

98 1John 3:2. This brief ninth section of chapter 6 is missing in its entirety from an abbreviation of Peter's work that circulated in manuscripts especially in monasteries in the area of southern Germany and Austria. Perhaps the earliest example of this line of abbreviated text without chapt. 6.9, and the earliest dated copy of Peter's treatise known to me, is Sigmaringen, Fürstlich Hohenzollern'sche Bibliothek MS. 12 (anno 1303).

99 Cf. Roger Bacon, *Persp.*, 3.1.3, ed. and trans. Lindberg, 262. At this point in his treatment of optics, Bacon is writing of plane mirrors and observes that "turres videntur in aqua everse, cum fiat reflexio a plana superficie aque" (towers seem to be upside down in water since they are reflected from the plane surface of the water). Peter's statement (in *P7P8* and the printed text) that *altitudines uidentur euerse* (vertical dimensions seem to be upside down) to someone looking in a mirror may look back to Euclid, *De spec.*, proposition vii, ed. and trans. Takahashi,

So also if the eye of our contemplation were judging things according to the mirror of holy doctrine, it would say that those things which seem to be on the left, i.e., unfavorable, are on the right, i.e., favorable, and vice versa in what pertains to the usefulness of misfortune and the danger of prosperity. The eye of our contemplation would also maintain that those things which appear to be above are below and vice versa, i.e., that proud things are worthless and humble ones valuable. And thus it is said explicitly in Isaiah 43: "I will say to the north wind," i.e., to adversity, "'Give,' and to the south wind," i.e., prosperity, "'Do not be a hindrance,'"[100] since those things which seem to be unfavorable are favorable and bring us to God, and things which appear to be favorable are unfavorable and hinder us from God.

Thus, one also reads in Genesis 48 that when the patriarch Jacob was about to bless the sons of Joseph, he crossed his arms and placed his left hand on Manasseh, who was on his right, and his right hand on Ephraim, whom Joseph stationed on his left.[101] Therefore, let the rich of this world be fearful – they who rejoice in keeping to the right in the present world – lest they keep to the left on the day of judgment, and at that time also in the presence of God*. For Manasseh, who is interpreted to mean "oblivion,"[102] signifying the rich who are unmindful of God, was placed on the left. For God placed the poor on the right, but the immoral rich on the left. And thus it is said in Proverbs 3: "Length of days," i.e., for the poor who live longer than the rich, "is in his right hand, and in his left hand wealth and glory,"[103] i.e., for the rich and those established as glorious in honors. The same thing that Jacob signified by crossing* his hands was also signified by the angel's positioning of his feet in Revelation 10: "I saw an angel descending from heaven, and he placed his right foot over the sea, but his left one over the land."[104] By "the sea," one designates the flood of misfortunes, by "the land," the prosperity of the world; by the right foot prosperity is symbolized, by the left, misfortune. Thus, the angel placed his right foot over the sea to

---

134: "Altitudines et profunditates a planis speculis reverse videntur" (Heights and depths are seen reversed in plane mirrors). On the reversal of right and left in a mirror, see *De spec.*, proposition xix, 158–61.

100 Isa 43:6.

101 Cf. Gen 48:13–14.

102 Jerome, *Lib. interp. Hebr. nom.*, Num., ed. de Lagarde, CCSL 72:82. The Hebrew *menasheh* is derived from the root *nashah* (to forget) in Gen 41:51, though the form of the verb in the biblical passage may have been adapted to fit the etymology proposed by the text. See also below, chapt. 12.2, p. 150.

103 Prov 3:16.

104 Rev 10:1–2.

show that worldly misfortunes are a matter of good fortune, but he placed his left foot over the land in order to signify that what the world has considered favorable is unfavorable.

The same thing is shown to us in the image of Jesus on the cross, which is raised up in churches like a book for the laity.[105] For he stretches his right hand to the north wind, while he extends his left hand to the south wind. Now, prosperity is signified by the south wind; misfortune by the north wind. In this way one ought to remark that the image of the Blessed Virgin has been placed on the side of the north wind under the right hand of Jesus on the cross, since the Blessed Virgin was exposed to misfortunes in this present life. And for this reason the just Simeon said to her, "His sword will transfix your soul," Luke 2.[106] We know, however, that she was loved above all others by God and has now been raised up to God's right hand. For the Lord exposes those dear to him to misfortunes and afflictions in this world when he has determined to place them on the right hand of his glory at the end of time.

So, therefore, when we turn our eye to the mirror of Holy Scriptures, the right seems to us the left and vice versa. Likewise, what is thought to be below will appear to be above and also vice versa. For Holy Scriptures bless the poor and raise them up, Matthew 5: "Blessed are the poor in spirit,"[107] and so on, but Scriptures call the powerful and the rich pitiful and humbled: "Go now, you who are rich; weep and wail in your miseries," James 5.[108] He had his attention turned fully to that mirror who said, "I saw servants riding on horses and princes walking on the ground."[109] For those who are held to be princes and lords are, in fact, slaves, subjected to miseries and delivered up to calamities. In book 7 of the *Memorable Deeds*, Valerius Maximus speaks of a certain king of exacting judgment. After the diadem had been given to him and before he placed it on his head, he considered it for a long time and said, "Oh, head band, more renowned than auspicious! If someone fully knew from how many cares, dangers, and torments he would have to recover, he would surely not want to pick it up from where it was lying on the ground."[110]

105 In the *Scholastica historia: Historia evangelica*, 5 (PL 198:1540A), Peter Comestor notes that church paintings are like books for the laity (*libri laicorum*). Peter echoes this phrase here.

106 Luke 2:35.

107 Matt 5:3.

108 Jas 5:1.

109 Qoh 10:7.

110 Valerius Maximus, *Factorum et dictorum memorabilium libri IX*, 7.2.Ext.5, ed. John Briscoe (Stuttgart and Leipzig, 1998), vol. 2, 446.

Quite rightly did he judge something to be of lesser value which seems to many people today to be of greater worth.

On the 11th extraordinary phenomenon. That knowing how to differentiate sins is necessary for prelates.*

It has been proven in the science I have frequently mentioned that the size of an object cannot be verified when the object is perceived through refracted rays of light, yet it can be if it is seen in direct lines, and namely this concerns any object that is seen at one time in the air and at another in water.[111]

In similar fashion, a sin can be comprehended with certainty according to the degree of its own proper magnitude by someone who looks at the sin directly with the eye of reason. And this is the way a professor or anyone else who is studious looks at sin: by reflecting on and inquiring into the truth through an examination of each and every thing that ought to be known about the degrees of sins. And indeed this concept is necessary above all for prelates, who are required to have the knowledge to differentiate "between different kinds of lepers," as it is said in Leviticus 13.[112] Now, when a sinner commits a sin, he does not see the fault directly yet, but rather through a refracted line, or an oblique line (or obliquely),[113] for he does not perceive the sin's deformity or evil, but rather the pleasure connected with it, since according to Dionysius no one does what he does with an eye to evil.[114] For this reason Aristotle says in the *Ethics* that every evil person is lacking in knowledge, namely since when he sins, he errs in making a choice.[115] And in Proverbs 14: "They err who

111 Cf. Roger Bacon, *Persp.*, 3.2.4, ed. and trans. Lindberg, 312–14.

112 Deut 17:8; cf. Lev 13:2–46.

113 The reading in *P1* (*obliquam vel oblique*) is the result of scribal dittography or an attempt to provide two possible expansions of an abbreviation. *P7P8* reads simply *oblique* at this point.

114 Cf. Ps.-Dionysius the Areopagite, *De divinis nominibus*, 4.19, trans. Robert Grosseteste et al., ed. Phillipe Chevallier et al., in *Dionysiaca: Recueil donnant l'ensemble des traductions latines des ouvrages attribués au Denys de l'Aréopage* ... ([Bruges], 1937), vol. 1, 236. For the Greek text, see the edition by Beate Regina Suchla, Corpus Dionysiacum, 1; Patristische Texte und Studien, 33 (Berlin and New York, 1990), 163.

115 Cf. Aristotle, *Ethica Nicomachea*, 3.2, trans. Robert Grosseteste (textus recognitus), ed. René Antoine Gauthier, in *Ethica Nicomachea*, Aristoteles Latinus, 26.1–3; Corpus Philosophorum Medii Aevi, 1.26.1–3/1–5 (Leiden and Brussels, 1974), Fasc. 4, 412.

perform evil."[116] And because of this fact it sometimes happens that while he is lecturing or making a determination about the magnitude of a sin, a professor will judge most correctly, because he is looking at the sin directly then, but afterwards when temptation attacks him he will succumb and will make an error in judgment by choosing to do what he had condemned earlier while teaching, and it happens in this way because instead of seeing the sin as he had earlier, he perceives it indirectly at that time.

If only our theologians were as judicious in making choices when they measure the quantities of faults as they are in making determinations when speculating on them! Since they are not so judicious, this is possibly the reason why they are found to be lazy in taking action, although they are impetuous in learning. A fool refuted a crowd of these theologians very nicely once at Paris, for when a number of theologians had gathered together in some of the schools in Paris, a fool came in and said, "I ask all of you which one is better: for someone to put into practice what he knows, or for him to learn what he doesn't know?" While they disputed the question he had just put to them and argued *pro* and *contra*, the fool remained silent, listening to their altercation and waiting to see where it would end. At last they came to a conclusion and proved that it is better for someone to do what he already knows than to learn what he does not know, since as the apostle Paul says in Romans 2: "Not those who only hear the law, but those who perform it will be justified."[117] And Isidore says in *On the Highest Good*: "A person who reads properly will be quicker to put into practice what he might read than he is to gain knowledge, for it is a lesser sin not to know what you should desire to know than not to carry out what you know."[118] "Thus," said the fool, "you're all crazy, you who labor day and night only to learn what you don't yet know while you don't care to put in practice what you do know."[119]

On the 12th extraordinary phenomenon, from which it is deduced that the judgment of a person's sight is deceived concerning the rich and the poor.*

Moreover, it is proven in Perspectivist science that a visible object appears to be larger than it actually is when the eye is in a medium of lesser

116 Prov 14:22.

117 Rom 2:13.

118 Isidore of Seville, *Sententiae*, 3.8.6, ed. Pierre Cazier, CCSL 111 (Turnhout, 1998), 229–30 (PL 83:680A).

119 Cf. Tubach, *Index exemplorum*, no. 4201.

density, and the reverse happens when the eye is located in a medium of greater density. For this reason a visible object in water will necessarily appear larger when the eye is in the air, but it will appear smaller when the eye is submerged in water and the object is situated in the air.[120]

In similar fashion it often happens that when a poor person living in the dry land of poverty sees someone overflowing with worldly riches, he will consider the rich man to be great, and for this reason the Psalm says: "They have called people blessed who own such things."[121] But he is deceived in the judgment of his sight, and so a certain philosopher says: None of those who are exalted because of their rank are great because of this fact, but they appear great to you because you are measuring them along with their pedestal. And neither is a grain of millet great just because it has been placed on top of a mountain.[122] And Bernard to Eugene: "Because you are the supreme pope, are you therefore supreme?"[123] On the other hand, when someone who is immersed in the transitory riches of this world sees a poor person separated by a great distance from worldly riches, he considers him to be of ordinary size, when in fact he is great to the divine eye that perceives the poor person and does not err in judgment. For this reason it was said about John the Baptist: "He will be great in the eyes of the Lord," Luke 1.[124] In the eyes of the rich, poor people are held to be very small, so they could say what the spies said in Numbers 13: "We saw the sons of Anak there, compared to whom we feel like locusts."[125]

Note that when master Alan, an outstanding professor and still poor, had been invited to lunch by one of his former students who had already been made a bishop, the man observed Alan's poverty and said, "I'm not a little surprised, master, that your students have become important men – one is an abbot, another a bishop, another an archbishop – but you've been abandoned in the ludicrous state of poverty." As an honest and judicious person, Alan felt just the contrary and is reported to have given the following response: "You don't understand the acme of a most perfect worthiness and a person's true magnitude," he said. "For it's not great to be a bishop, but rather a good cleric. The proof: With the vote of three

120 Cf. Roger Bacon, *Persp.*, 3.2.2, ed. and trans. Lindberg, 294.

121 Ps 143:15.

122 Cf. Seneca, *Ad Lucilium epistolae morales*, 76.31, ed. Leighton D. Reynolds (Oxford, 1965; reprint 1987), vol. 1, 243. Seneca, however, writes of those who have been elevated by riches and honors, not by rank (*nomen*, in *Pr1*).

123 Bernard of Clairvaux, *Tract. de consid.*, 2.7.14, ed. Leclercq and Rochais, in *Opera*, vol. 3, 422.

124 Luke 1:15.

125 Num 13:34.

decadent canons to whom the power of election is given, one man will become bishop, but if all the saints in heaven and all the people on the earth were to say before God with one heart and one voice, 'Martin is a good cleric,' that wouldn't be reason enough for Martin to be a good cleric if he still continued in his folly." For indeed this bishop, whose eye, as it were, had been immersed in the denser medium of riches, considered Alan ordinary (though he was truly great), when he observed him in the thinner medium of poverty. It is also reported that Alan added afterwards: "I'll prove to you, lord bishop, that a poor person is the king and lord of this world, for the world doesn't obey the rich person's will in all matters, however great his power may be, for he fears to lose what he has and at the same time he wants to get what he doesn't have, and in this way the world is opposed to his will, since he wants to make it impossible to lose what he possesses and to have acquired what he doesn't possess already. But the pauper possesses the world, which is fully responsive to his will, because since he hates riches, he hopes to acquire nothing, and since he possesses nothing, he fears to lose nothing."[126] And in this way what Fortunatus says is true:

> The poor man in his straitened condition rules when he possesses God.[127]

And Seneca, speaking about Diogenes, the greatest hater of riches, who responded as follows to King Alexander when he offered him gifts: "Take them back," he said, "they're not gifts for human beings, but burdens for cattle"[128] – about this Diogenes, I say, Seneca says elegantly: "Diogenes was more powerful than Alexander of Macedonia, who possessed all things. For what Diogenes did not want to acquire was greater than what Alexander was able to give."[129]

Nevertheless, today this very deception of sight has so invaded the minds of the powerful and the rich that among them a person is held to be nothing unless he either is rich or at least is thought to be rich accord-

126 For the dialogue between Alan and his student, cf. Tubach, *Index exemplorum*, no. 3844. The Alan referred to here is presumably the theologian and poet, Alan of Lille (d. 1202). See above, chapt. 6.8, p. 33.

127 Venantius Fortunatus, *Carmina*, 8.3.296, ed. Friedrich Leo, in *Opera poetica*, MGH, Auctores antiquissimi, 4.1 (Berlin, 1881), 189 (cf. Walther, *Proverbia*, no. 20925).

128 Cf. Tubach, *Index exemplorum*, no. 1673.

129 Vincent of Beauvais, *Spec. doct.*, 4.104, 4.148, in *Speculum quadruplex*, vol. 2, 358c, 386a. Cf. Seneca, *De beneficiis libri vii*, 5.6.1, ed. Karl Hosius, in *De beneficiis libri vii, De clementia libri ii*, 2nd ed. (Leipzig, 1914), 121.

ing to his appearance. Costabulus records that after a certain philosopher was turned away from coming before a king, he considered the famous poetic line:

> The senate is barred to the poor; wealth gives honors.[130]

Then he dressed in the finest clothing and was admitted – as soon as he arrived, the door stood open to him. Having entered, he began to kiss his own clothing reverently and surprisingly before the king. When the king marveled about this and asked him why he behaved so, the philosopher responded, "I honor what is honoring me, since what virtue was unable to get, clothing has obtained. Oh, vanity of vanities," he said, "more is granted to vestments than to virtues, more to beauty than to honesty!"[131] It is written in the *Deeds of the Romans* that the Romans, who used to deify great men, argued in their senate whether Jesus ought to be counted among the number of the gods since he had brought about so many miracles, and at last it was decided that he should not be accepted as a god since he would have no worshipers, for he preached poverty which everyone naturally despises.[132]

On the 13th extraordinary phenomenon. And on preserving humility in a lofty position.*

Alhacen* teaches in book 7 of his *Optics* that stars located near the horizon to the west or the east appear larger to the eye than when they have risen to the middle of the sky.[133]

130 Ovid, *Amores*, 3.8.55, ed. Edwin J. Kenney, in *Amores, Medicamina faciei femineae, Ars amatoria, Remedia amoris*, 2nd ed. (Oxford, 1994), 94 (cf. Walther, *Proverbia*, no. 4763).

131 Cf. Tubach, *Index exemplorum*, no. 1113. Cf. Pope Innocent III, *De miseria conditionis humane*, 2.39, ed. and trans. Robert E. Lewis (Athens, GA, 1978), 197–99. The *exemplum* was also used by Stephen of Bourbon; see Jacques Berlioz, "Etienne de Bourbon, *Tractatus de diversis materiis predicabilibus*," in Jacques Berlioz and Marie Anne Polo de Beaulieu, *Les Exempla médiévaux* (Carcassonne, n.d.), 148. Peter attributes the authorship of the narrative to the ninth-century Arabic writer Costa ben Luca, known frequently in Latin as Constabulus or Costabulus; see Hans H. Lauer, "Qusta ibn Luqa," *Lexikon des Mittelalters*, 7 (Munich and Zurich, 1995), 378. The *exemplum* also inspired an English reader to record a bit of Middle English verse at this point in the copy of Peter's text that he was reading. See Richard Newhauser, "A Middle English Poem on the Fleeting Nature of Material Wealth," *Medium Ævum* 71,1 (2002): 74–81.

132 Cf. Tubach, *Index exemplorum*, no. 1008.

133 Cf. Alhacen, *De aspect.*, 7.7 [7.70–71], ed. and trans. A. Mark Smith, in *Alhacen on*

In similar fashion, the greater the heights of honors which saintly people appear to ascend, the more they denigrate themselves through humility and the more inferior they desire to appear in the eyes of other people, according to the verse in Ecclesiasticus 3: "To the degree to which you are great, humble yourself in all things."[134] As it can be said about such people in an elegant metaphor that was said about Esther: "A small spring grew into a river, and it was turned into the sun and the moon."[135] Namely, even if such people are great in honor, as long as they are not proud because of it, they are small in humility, and for that reason they are metaphorically turned into the sun and the moon, since when they are elevated like the celestial luminaries they become more humble. The glorious Virgin was this kind of person: Although she had been raised to the height of such great honor as to become the mother of God, she immediately showed that she was small, like a star, and dedicated herself as a handmaiden. For this reason we fittingly call her the "star of the sea."[136] Concerning a man named Constantine, who was very small in stature but large in virtue, Gregory relates this in the first book of the *Dialogues*: When a certain man who had arrived in order to meet Con-

---

Refraction. A Critical Edition, with English Translation and Commentary, of Book 7 of Alhacen's De Aspectibus, the Medieval Latin Version of Ibn al-Haytham's Kitab al-Manazir, Transactions of the American Philosophical Society, 100, Part 3, sections 1–2 (Philadelphia, 2010), vol. 1, 141–43 (Latin text), vol. 2, 328–30 (English translation). Cf. also Roger Bacon, *Persp.*, 2.3.6, ed. and trans. Lindberg, 226, who refers to both Alhacen and Ptolemy's *Optica* (and see Lindberg's analysis of Bacon and Ptolemy on this optical phenomenon in *Roger Bacon and the Origins of Perspectiva in the Middle Ages*, 378n478). The visual phenomenon by which celestial bodies appear larger near the horizon than they do at higher altitudes in the sky is referred to as the "moon illusion" and is discussed at length on the basis of its treatment by Ptolemy and Alhacen in Abdelhamid I. Sabra, "Psychology versus Mathematics: Ptolemy and Alhazen on the Moon Illusion," in Edward Grant and John E. Murdoch, eds., *Mathematics and Its Applications to Science and Natural Philosophy in the Middle Ages. Essays in Honor of Marshall Clagett* (Cambridge, Engl., 1987), 217–47.

134 Sir 3:21.

135 Esth 10:6.

136 Mary was associated with the north star as early as the hymn *Ave maris stella*, preserved earliest in a manuscript of the ninth century. The text is edited in *Cantiones Bohemicae*, ed. Guido Maria Dreves, Analecta hymnica medii aevi, 1 (Leipzig, 1886), 49 (no. 4). See Heinrich Lausberg, "Ave maris stella," in *Lexikon für Theologie und Kirche*, 1, 2nd ed. (Freiburg i. Br., 1957), 1141–42; Herbert Musurillo, "The Mediaeval Hymn, *Alma Redemptoris*: A Linguistic Analysis," *The Classical Journal* 52,4 (1957): 171–74.

stantine said to him with contempt that there was nothing honorable about him, immediately the man of God was exceptionally happy, fell into his embrace, and with the most fervent love began to hug him and to kiss him sweetly, giving him thanks for having judged him in this way. "You," he said, "are the only one who's seen me with open eyes."[137] Constantine had quite clearly ascended to a state of lofty merit and wanted to appear small in the eyes of other people, in the manner of a star in the heavens.

We consider it extraordinary that when a star ascends to the middle of the sky, it appears smaller, although to the eye of the observer it is ascending higher,[138] and in the same way we consider it extraordinary that the more exalted someone is in honor, the more humble he wants to appear, since this is a rare occurrence. For it rarely happens that a degree of honor does not appear as a swelling of the heart in a ruler, for according to Bernard: "It is not great to be humble in a state of abjection; humil-

137 Cf. Gregory the Great, *Dial.*, 1.5.4–5, trans. Antin, ed. de Vogüé, SC 260:60–63 (PL 77:180C), who names the protagonist of the narrative "Constantius," not "Constantine." Cf. Tubach, *Index exemplorum*, nos. 1221, 5016.

138 One can note that the archetype of Peter's text contains an error here. The manuscripts representing the most authoritative line of transmission read at this point (following *P7*): "Et sicut reputamus mirabile quod stella, cum ascenderit ad celi medium, maior appareat, cum iam ad intuentis oculum non magis accedat, sic eciam ..." (We count it extraordinary that when a star ascends to the middle of the sky, it appears larger, although at the same time it is not drawing closer to the eye of the observer, and in just the same way ...). This, however, contradicts the scientific description of the moon illusion that Peter had just given at the beginning of chapt. 6.13, though it anticipates the moral lesson in humility that is to be drawn from this optical phenomenon. Some scribes who paid attention to the scientific error undertook changes in the text to make it more scientifically accurate, as apparently the scribe of the copy text represented in *Pr1* did (which reads: "Et sicut reputamus mirabile quod stella, cum ascenderit ad celi medium, minor appareat, cum tamen ad intuentis oculum magis ascendat, sic eciam ..."). In *P8*, the reading of the authoritative transmission is also present, but was later altered, so that *maior* has been corrected to *minor*, *ad intuentis oculum* to *ab intuentis oculo*, and *accedat* to *recedat*, to yield the following: "Et sicut reputamus mirabile quod stella, cum ascenderit ad celi medium, minor appareat, cum iam ab intuentis oculo non magis recedat, sic eciam ..." (We count it extraordinary that when a star ascends to the middle of the sky, it appears smaller, although at the time it is not moving any further from the eye of the observer, and in just the same way ...). The text of *Pr1*, then, takes its place among many other attempts to deal with this crux by revising it to be scientifically accurate.

ity is precisely a great and rare virtue when it is being honored."[139] And as Gregory says: It is great not to desire honor, it is even greater to reject it once it has been offered, but the greatest is to make a habit out of not being proud.[140]

139 Bernard of Clairvaux, *Homiliae in laudibus Virginis matris*, 4.9, ed. Jean Leclercq and Henri M. Rochais, in *Opera* (Rome, 1966), vol. 4, 55.

140 Cf. William Peraldus, *Summa de vitiis*, 6.3.39, in *Summa virtutum ac vitiorum Guilhelmi Paraldi Episcopi Lugdunensis de Ordine Predicatorum* (Paris, 1512), vol. 2, fol. 187va, who attributes the following to Gregory: "Aliquantulum facile est humanam laudem non appetere, difficillimum oblatam respuere" (It is a small and easy matter not to desire people's praise; it is very difficult to reject it once it has been offered).

CHAPTER SEVEN

On Moral Instruction According to Twelve Properties That Have Been Ascertained in the Corporeal Eye

Now, the eye is doubled in number; each one being exactly like its companion; varied in color; spherical in shape; covered by eyelids; located in the head; receptive to visible *species*; able to perceive another object, but not itself; able to gain knowledge of visible things in different ways according to variations in their distance; able to establish the direction for the entire body; indicative of the secrets of the mind; and drowsy after being awake for a long time.

On the value of the communal life.*

In the first place, then, the eye is doubled in number, and this is so, as Alhacen* testifies, so that the one might be strengthened by the other, so that a defect in the one might be made up by the other.[1] And for this reason Bernard Silvestris, speaking in metrics about eyes in book 2 of the *Cosmographia*, the *Microcosmos*, says:

> Not in vain are there two of them, for if one grows weary,
> The other will take its place to support the work of its partner.[2]

1 On the reason for doubling the eyes, see Roger Bacon, *Persp.*, 1.4.4, ed. and trans. Lindberg, 56, who paraphrases Alhacen, *De aspect.*, 1.8 [7.9], ed. and trans. Smith, in *Alhacen's Theory of Visual Perception*, vol. 1, 70 (Latin text), vol. 2, 388 (English translation). Both authors attribute to the beneficence of the Creator the presence of two eyes so that if one is injured the other remains operational, and they also mention that two eyes make the face more beautiful. They do not refer to the strengthening of the remaining eye when one is injured. The doubling of the eyes is mentioned again below in chapt. 12.2, p. 152.

2 Bernard Silvestris, *Cosmographia*, 2.14.37–38, ed. Peter Dronke, Textus Minores, 53 (Leiden, 1978), 151 (cf. Hans Walther, *Initia carminum ac versuum medii aevi posterioris latinorum*, Carmina medii aevi posterioris latina, 1 [Göttingen, 1959], no. 12118).

In this fact, nature commends a communal life to us by its own example. For just as the eyes strengthen each other, since they are in a single head, likewise many people residing together in a community give each other mutual support. And therefore Solomon says in Ecclesiastes 4: "Two together are better than one, for they have the advantage of each other's fellowship."[3] Namely, they assist each other in the task of abandoning evil. Therefore he adds: "If one falls, he will be helped up by the other. Woe to the solitary person, since if he falls, he has no one to support him!"[4] About this matter Jerome relates this *exemplum* to Rusticus: "I saw a Greek youth," he says, "in a certain coenobitic monastery who was unable to extinguish the burning of his flesh by abstinence or by any amount of work. The abbot of the monastery saved this youth, who was being sorely tested, using this strategy: He ordered a certain man to attack this youth with sticks and words of abuse and, after he had inflicted these injuries, to be the first to take up a formal complaint against the youth. The witnesses who were called spoke on behalf of the man who had brought the complaint. The young monk began to cry because of this falsehood. No one stood up for him; only the abbot took pains to skillfully set out his defense, 'lest the brother be overcome by a too abundant sadness.'[5] What more can I say? A year later when the youth was questioned about whether his former thoughts were still molesting him, he said, 'Father, if I have no right to live, how could fornicating please me?' Oh," says Jerome, "if this youth had been alone, with whose help would he have overcome temptation?"[6] Therefore Seneca says: "A large number of sins can be eliminated if a witness stands by those who are about to sin."[7]

The communal life is also useful for sustaining someone in the good, Proverbs 18: "A brother who is helped by his brother is like an unyielding city."[8] One reads in the *Sayings of the Fathers* that a certain man who was striving earnestly for salvation consulted an elder concerning which of the two kinds of life would be the greater benefit for him to follow, namely, the eremitic or the coenobitic*. The elder did not want to respond to him in words, because he didn't want to expose himself to accusation, but he answered with an example. He ordered the man to bring him a jar filled with burning coals. The man obeyed his commands

3 Qoh 4:9.

4 Qoh 4:10.

5 2Cor 2:7.

6 Jerome, *Epistulae*, 125.13.1–2, ed. Isidor Hilberg, CSEL 56 (Vienna and Leipzig, 1918), 132; cf. *Vitae patrum*, 3.220, PL 73:810B–C.

7 Seneca, *Ep. mor.*, 11.9, ed. Reynolds, vol. 1, 25.

8 Prov 18:19.

and when he brought it, the elder took out the coal which he saw was burning brightest and set it aside on the ground where it died out after a little while, since it was by itself. The man did not understand this and asked the elder what he was driving at: "I beg you, what've you done?" And the elder said, "If you had paid attention, your question would've been answered already. Look, the piece of coal that formerly burned brighter than all the others when it was situated among them quickly died out when it was surrounded by the cold air, since it was then separated from the others. The rest continue to glow even now because they keep themselves warm with the heat of mutual fellowship. Thus, a person who remains in a holy community is kept warm in what is morally good, but living alone or surrounded by the fellowship of perverse people, who've been made cold in sin, he easily becomes lukewarm and the fire of brotherly love dies out in him."[9] This is also what is said in Ecclesiastes 4: "If two sleep in the same bed, they warm each other mutually. How can one keep himself warm?"[10]

Thus, the fact that the eye is doubled recommends to us a holy fellowship and a mutual love. And therefore, the Lord sent out disciples, so to speak the church's spiritual eyes, "two by two," Luke 10.[11]

On the second moral instruction, namely on fools saying many things and doing few, knowing a great deal and loving only a little.*

Second, an eye should be absolutely similar to its companion, for it is monstrous if one eye is larger than the other or is differently colored. Nevertheless, we can see this monstrosity today in many people, for although we have twofold eyes, namely that of the intellect and that of the affections, in some people these eyes are dissimilar in color – they teach one thing and behave in a different way –, while in other people these eyes are unequal in size, for they say many things and do few, they know a great deal and love only a little: That is to say, they pay a great amount of attention to enlightening their intellect and care too little about kindling their affections. Truth illuminates the eye of the intellect, brotherly love the eye of the affections. David asked that both of these be

9 Cf. Bonaventure of Balneoreggio, *Sermo 36 (de Sancto Nicolao)*, 9, in Saint Bonaventure, *Sermons de diversis*, ed. Jacques Guy Bougerol (Paris, 1993), vol. 2, 483.

10 Qoh 4:11.

11 Luke 10:1.

illuminated equally when he said, "Give light to my eyes."[12] But in many people the left eye is illuminated a great deal while the right one has been blinded, which one reads was prophesied in Zechariah 11 concerning the prelates of the church: "Oh shepherd and idol, forsaking the flock, his arm will wither in dryness and his right eye, growing dark, will be blinded."[13] The devil seeks eagerly to pluck out humanity's right eye rather than the left one since he endeavors to blind a human being with respect to spiritual matters, for he knows that if he were to tear out a person's right eye, he would render him useless for spiritual battle. And this has been symbolized for us in 1 Samuel 11 where one reads that "the besieged men of Jabesh said to Nahash the Ammonite, 'Take us as allies and we will serve you.' And Nahash responded, 'On this condition will I make a treaty with you, that I gouge out all of your right eyes.'"[14] Nahash is interpreted as "serpent," Ammonite as "oppressing" or "afflicting," and he represents humankind's enemy who is a serpent alluring us with suggestions, oppressing us with the stings of conscience, afflicting us by the exacting of punishment.[15] The serpent wishes to sign this treaty with us so that after we have lost our left eye, that is, a knowledge of temporal matters, he can pluck out the right eye, by which the desire for eternal matters is signified.

Oh, how many are there today who see clearly in temporal matters and are blind when it comes to divine ones, who exert themselves greatly when it comes to scholastic questions, but just moderately when it comes to morals! About them Job 24 can be said: "He afflicts the barren woman who does not give birth, and he has not helped the widow."[16] By the barren woman one understands the intellective faculty, which does not give birth, i.e., is not deserving, since all merit consists in the use of the will; by the widow one understands the affective faculty, which is widowed in many people, since it has not been joined to its heavenly bridegroom through desire.[17] People of this sort are monstrous, since one of their eyes, namely that of the intellect, is so large and wide that with it they consider carefully the heavens and the stars, and the courses of the planets and heavenly bodies, and the distance across the entire earth and

12 Ps 12:4.

13 Zech 11:17.

14 1Sam 11:1.

15 Cf. *Gloss. ord.*, 1Reg. 11:2, ed. Rusch, vol. 2, 17b in margin (PL 113:552D).

16 Job 24:21.

17 On the distinction between the intellective and affective faculties, see for example Hugh of Saint-Victor, *In Salomonis Ecclesiasten Homiliae xix*, 2, PL 175:141C, where the substance of the rational soul is divided into the two parts of cognition and affect. This dichotomy is typical of Hugh's thought.

regions of the world, so that one can say about them the verse in Zechariah 5: "Their eyes are through the whole earth."[18] But their other eye, namely that of the affections, is very small and scarcely anything at all, since they are too little moved to God. Such people can say that verse in Job 16: "My eye drips slowly to God."[19] For such people are very attentive to physical bodies and transitory matters, but they give too little thought to spiritual and eternal matters, and for this reason their eye copiously sheds tears to the world, but barely drips to God, since they are moved very little to God. Such people are like Jacob, since he preferred the love of Rachel to the love of Leah.[20] Rachel is interpreted to mean "seeing the beginning,"[21] and she stands for the speculative sciences that inquire into the origins of things and the foundations of astronomical conjunctions. Leah, on the other hand, is taken to mean "laborious,"[22] and she stands for the exercise of virtue, for virtue is concerned with what is difficult and arduous.

Let the sinfully curious hear what Seneca would say: But why do you torment yourself with a question which it is more useful to have considered with disdain than to have studied?[23] And the same author says: "Even if one had a great amount of time, one would have to spend it sparingly in order to have a sufficient amount left for necessary things. But as it is now, what madness is it to inquire into superfluous matters when time is so scant?"[24] And Hugh of Saint-Victor in the *Treatise on the Formation of Novices*: "In scriptural readings you should seek whatever instructs your morals in virtue rather than what might whet your appetite for subtlety, and you should want to be taught by the precepts of Scriptures more than to be entangled in scholastic questions."[25] Concerning Saint Edmund one reads that his mother appeared to him and reproved him because he was giving his attention to mathematical figures, and on her right hand she drew three circles, one under the other, in the first of which was written "Father," in the second "Son," and in the third "Holy Spirit." "Dearest son," she said, "give your attention to these figures." Educated by what he had seen, he abandoned the overly curious sciences and rushed to study Holy

18 Zech 5:6.

19 Job 16:21.

20 Cf. Gen 29:30.

21 Jerome, *Lib. interp. Hebr. nom.*, Gen., ed. de Lagarde, CCSL 72:70.

22 Ibid., 68.

23 Cf. Seneca, *Ep. mor.*, 49.6, ed. Reynolds, vol. 1, 129.

24 Ibid., 48.12, ed. Reynolds, vol. 1, 127.

25 Hugh of Saint-Victor, *Liber de institutione novitiorum*, 8, PL 176:933D.

Scriptures.[26] Speaking about the mathematical sciences in the fifth book of the *Confessions*, Augustine says on this matter: "Unhappy is the person who knows those things, but does not know you, but blessed is he who knows you, even if he has no knowledge of those things. However, if someone knows both you and those matters, he is not more blessed because of them, but rather he is blessed only because of you, if he glorifies you as God and gives thanks to you and does not dissipate himself in his imaginings. For someone who knows that he owns a tree and gives thanks to you for the use of it, even though he does not know either how many cubits high it is or how wide it spreads, is better than someone who measures it and counts all its branches, but neither owns it nor knows or has learned about its Creator. And it is just the same for someone with faith, whose wealth is the whole world and who, possessing nothing, owns everything as it were by clinging to you whom all things serve, although he does not even know the course of the Big Dipper. It is foolish to doubt that he is surely better than one who measures the heavens and counts the stars and weighs the elements, neglecting you 'who have arranged all things by number and measure and weight.'"[27]

The third moral instruction. On maintaining the memory of death, which worldly people hate.*

Third, different people have eyes that are differently colored, and the operation of the visual power changes with the different colors. For this reason Aristotle says in book 19 of *On Animals* that those with black eyes are sharp-sighted during the day, but can see only weakly at night, and those with bluish grey eyes see more clearly at night than during the day, as is observed in cats and other animals with bluish grey eyes.[28] Applying

26 *Vita b. Edmundi*, 16, ed. Martène and Durand, in *Thesaurus novus anecdotorum*, vol. 3, 1787D–E (cf. Tubach, *Index exemplorum*, no. 1068). Edmund Rich (of Abingdon; 1180–1240) studied in Oxford and Paris and later became archbishop of Canterbury. Illustrations of this episode often depict the trinitarian diagram as three interlocking circles.

27 Wis 11:21. The passage is taken from Augustine of Hippo, *Conf.*, 5.4.7, ed. Verheijen, CCSL 27:60.

28 Cf. Avicenna, *De animalibus*, 19, in *Opera in lucem redacta ac nuper quantum ars niti potuit per canonicos emendata* (Venice, 1508; reprint Frankfurt/M, 1961), 63vb; cf. Aristotle, *De generatione animalium*, 5.1, trans. William of Moerbeke, ed. Hendrik J. Drossaart Lulofs, Aristoteles Latinus, 17.2.5; Corpus Philosophorum Medii Aevi, 1.17.2.5 (Bruges and Paris, 1966), 158; and trans. Michael Scot, ed. Aafke M.I. van Oppenraaij, in *De animalibus. Michael Scot's Arabic-Latin Translation. Part Three: Books XV–XIX: Generation of Animals*, Aristoteles Semitico-latinus, 5 (Leiden, New York, and Cologne, 1992), 216–17.

this matter spiritually, one can say that because bluish grey eyes have a weak spirit and a limited amount of moisture, they signify those who are wise in the ways of the world, in whom the moisture of grace is deficient and the power to do good deeds is only tepid. And such people perceive more clearly at night than during the day since they turn all of their senses to worldly matters, and in this they are like the mole which is wise and sagacious under the earth and knows many winding by-ways there; however, when it comes up above ground, it acts foolishly and loses all power of judgment.[29] So, too, the wise of this world are sagacious in earthly and temporal matters, but in spiritual ones they are sluggish and ignorant. On the other hand, spiritually speaking, those people whose eyes have been firmly stamped by a reminder of death have mental eyes with a black iris. And these people are not able to see by night, but they can during the day, since such people do not see nor do they approve of shady behavior, but rather deeds of light, i.e., virtuous deeds, Ecclesiasticus 7: "Keep in mind your last days and you will not sin."[30] The bridegroom is praised for this spiritual and laudatory blackness of mind in the Song of Songs where it is said: "His hair is black like the crow."[31] Thoughts are signified here by hairs because they are delicate and, as it were, innumerable. Thus, the hair on the head is black, spiritually speaking, when there is a continual contemplation of death in the mind.

One reads that the ancients approved of this reminder of death in their words as well as in their deeds. For this reason Seneca says in his *Letters*: Death waits for you everywhere; you, too, if you would be wise, will expect death in every place.[32] And the same author says in *On Morals*: "Unexpected death prevents many people from changing their life, and thus each day ought to be organized as if it were the last."[33] And so the poet Horace says:

> Believe that every day that has dawned is your last.[34]

29 Cf. Tubach, *Index exemplorum*, no. 3315.

30 Sir 7:40.

31 Cant 5:11.

32 Cf. Seneca, *Ep. mor.*, 26.7, ed. Reynolds, vol. 1, 76.

33 Ps.-Seneca, *De moribus liber*, 10, ed. Friedrich Haase, in *Opera quae supersunt* (Leipzig, 1853), vol. 3, 462; reprint in *Opera quae supersunt. Supplementum* (Leipzig, 1902), 60. For *differre* = "to change," see Charles du Fresne (Sieur du Cange), *Glossarium mediae et infimae latinitatis*, ed. nova (Niort, 1883–1887; reprint Graz, 1954), vol. 3, 2 (q.v. *differre*).

34 Horace, *Epistulae*, 1.4.13, ed. Edward C. Wickham, 2nd ed., ed. Heathcote W. Garrod, in *Opera* (Oxford, 1901) (cf. Walther, *Proverbia*, no. 19884).

And someone else said:

> Always remember to keep in mind that it is necessary to die.[35]

And as Jerome says in book 1 of *Against Jovinian*: "When he was defining philosophy, Pythagoras said as follows, 'Philosophy is the contemplation of death striving daily to bring about the freedom of the soul from the prison of the body.'"[36] And as Jerome says, for the purpose of study Plato dedicated a country estate to the Academy on which earthquakes were a frequent occurrence, and for the following reason: so that the students, terrified by the earthquakes, would contemplate death and thus refrain from vices.[37] One also reads in the *Deeds of the Romans* that after the coronation of the emperor, and before he could issue any commands, those who built tombs would come before him carrying in their hands four kinds of marble of different colors, and in front of all the princes they would say to the emperor, "Of what kind of marble do you command that your tomb be made, emperor?" By this action they communicated two things to him: that he could not consign death to oblivion and that he should rule the empire with mercy.[38]

In fact, there is great wisdom in the remembrance of death. For human creation began from dust, namely since "God created man from the mud of the earth"[39] – in other words, God created humanity as dust –, and likewise at the end of time glorification will begin from dust when the dust of our bodies will be gathered together and will be re-formed into glorified bodies. And in the same way, spiritual renewal through grace begins from dust, i.e., the consideration of death, in which a human being returns to dust, according to the verse of the Psalm: "You take away their spirit and they die, and they return to dust again."[40] And since a reminder of death is so useful, therefore David, who wished that he would always have a mental eye open and alert to death, prayed saying, "Illuminate my eyes, lest I ever sleep in death; lest the peoples should ever say,"[41] and so on.

35 *Sententia communis* (Walther, *Proverbia*, no. 27941).

36 Jerome, *Epistula aduersus Rufinum*, 3.40, ed. Pierre Lardet, CCSL 79 (Turnhout, 1982), 110; cf. idem, *Adversus Iovinianum libri duo*, 2.9, PL 23:311D; cf. idem, *Ep.*, 127.6.1, ed. Hilberg, CSEL 56:150.

37 Cf. Jerome, *Adv. Iov.*, 2.9, PL 23:312A (cf. Tubach, *Index exemplorum*, no. 3819).

38 Cf. Tubach, *Index exemplorum*, no. 3410.

39 Gen 2:7.

40 Ps 103:29.

41 Ps 12:4; Ps 113B:2.

*The fourth moral instruction. On preserving a singularity of intention**.

Fourth, the eye is spherical or round. Now, a spherical shape is the simplest of all shapes, since it is completed by one line. Therefore, the simplicity of the shape of the eye exhorts us to a singularity of intention. The roundness of the shape also persuades us to direct the eye of our intention to an eternal reward in all that we do. This singularity of intention is indicated not only in the shape of the eye, but also in the method of seeing, for as it is said in the *Book on Sight*: "Whatever we see, we see in an angle such that the vertex is in the pupil of the eye, while the base reaches to the ends of the visualized object."[42] In this fact a correct intention is indicated that unites all the ends of our thoughts and affections with God, referring all things back to the praise of the Creator according to the passage in 1 Corinthians 10: "Whether you are eating or drinking, or anything else that you do, do everything for the glory of God."[43] Therefore, admonishing us to single-mindedness, the Lord said in Matthew 6, "If your eye is singular, your entire body will be full of light. If, however, it is vile," and so on.[44] That person has a singular eye who has the intention to honor God in everything he does; however, a person who intends something else shows himself to be unfaithful to the Lord. Therefore, this is fittingly added: "If, however, your eye is vile, your whole body will be full of darkness."[45] "He is, in fact, a faithful servant if he does not try to grasp in his own hands any of the abounding glory of his Lord which, though not originating in him, nevertheless passes through him."[46] So was Joab to his lord, for when he was fighting against Rabbah and the royal city was at the point of being captured, he sent for David, lest the victory be recorded under his name, as one reads in 2 Samuel 12.[47]

Many people, however, have an eye that is not singular, but rather twofold. This is like an old farmer's wife: If she raises her eyes to the heavens, she does so only out of love for her young chickens in order to keep a lookout for hawks. So it is with many*: Even if they devote themselves now and then to matters which appear to be heavenly, nevertheless they keep the eye of their intent focused on worldly matters. And so the

42 Euclid, *Liber de visu*, pref., ed. Wilfred R. Theisen, "*Liber de visu*: The Greco-Latin Translation of Euclid's *Optics*," *Mediaeval Studies* 41 (1979): 44–105, here 62.

43 1Cor 10:31.

44 Matt 6:22–23.

45 Matt 6:23.

46 Bernard of Clairvaux, *Sermones super Cantica canticorum*, 13.3.3, ed. Jean Leclercq, Charles H. Talbot, and Henri M. Rochais, in *Opera* (Rome, 1957), vol. 1, 70.

47 Cf. 2Sam 12:26–28.

eye of such people is not singular, but vile or evil. We see that even crossbowmen who want to visualize a straight path to the target or to shoot an arrow accurately with the crossbow are accustomed to close one eye and keep the other one open. So, too, someone who wants to perform his actions correctly uses one eye so that he directs his mind to the one thing that is necessary, that is to say to God, so that he can say with Augustine in the *Confessions*: "Lord, be our glory. If we are loved, let it be for your sake, so that your word may be feared in us. For no one who wants to be praised by human beings when you censure him can be defended by them when you judge him, nor is he saved when you condemn him."[48] As an indication of this fact, the church has developed the custom of singing "Glory be to the Father" at the end of every Psalm, as if we were to say for every one of our acts, "Not to us, Lord, not to us, but to your name give the glory."[49] For the excellent quality of a letter ought to be attributed to the writer, not to the quill. Therefore, since he carries out all things through us, to God alone belongs the honor and glory. Amen.

The fifth lesson. On avoiding vainglory.*

Fifth, the eye is covered by eyelids. By their nature, eyelids come in pairs for the purpose of guarding the eyes from external injuries and preventing the loss of the visual spirits.[50] As Aristotle says in book 4 of *On Animals*: Every animal without eyelids has weak eyesight, namely fish and hares whose eyes are always open, and therefore they experience an excessive loss of the spirits because of the lack of protection or the covering provided by the eyelids.[51]

In similar fashion, as long as they seek approbation from other people, those who do not hide their good qualities but show themselves off through vainglory and ostentation lose everything. For this reason Gregory says: "Whoever carries his treasure publicly on the road is asking

48 Augustine of Hippo, *Conf.*, 10.36.59, ed. Verheijen, CCSL 27:187.

49 Ps 113B:1.

50 For Alhacen, the visual spirits are the medium of transmission of the visible forms between the eye and the brain within the hollow optic nerve. See Alhacen, *De aspect.*, 1.8 [7.7], ed. and trans. Smith, in *Alhacen's Theory of Visual Perception*, vol. 1, 70 (Latin text), vol. 2, 388 (English translation).

51 Cf. Aristotle, *De historia animalium*, 4.10, trans. William of Moerbeke, ed. Pieter Beullens and Fernand Bossier, Aristoteles Latinus, 17.2.1.1 (Leiden, etc., 2000), 124; cf. idem, *Historia animalium*, 4.10 (537a3), ed. David M. Balme, vol. 1: Books I–X: Text, Cambridge Classical Texts and Commentaries, 38 (Cambridge, Engl., 2002), 204; cf. idem, *De partibus animalium libri iv*, 2.13 (658a4–5), ed. Bernhard Langkavel, Aristotelis opera, 1 (Leipzig, 1868), 50.

to be robbed."[52] And so it is said in the *Lives of the Fathers*: "Just as a treasure is quickly diminished after it has been discovered, so, too, once a virtue has been made public, no matter how, it perishes."[53] And a certain elder said, "Just as it cannot happen that a plant sprouts and bears fruit at one and the same time, so it is impossible for us to obtain the fruit of heaven and at the same time have worldly praise and admiration."[54] As an indication of this fact, Moses' hand continued to be healthy as long as he concealed it in the fold of his cloak, but when he revealed it publicly, it became leprous, Exodus 4.[55] One reads in Isaiah 39 that after Hezekiah, the king of Jerusalem, had received a sign of protection from the Lord through the hand of the prophet Isaiah (for the shadow of the sundial had gone back ten steps, although the shadow normally went forward, not backward),[56] the king of Babylon sent messengers to Hezekiah bearing letters and gifts. King Hezekiah was overjoyed at their arrival and showed them the storeroom of his spices, his gold and silver, and the entirety of his treasures. Isaiah was sent to Hezekiah on behalf of the Lord and predicted that all of these things would be carried off to Babylon.[57] The Babylonian messengers are explicitly said to have arrived with gifts after the Lord had granted a miracle to the king, since each person is particularly vexed by temptations at the instigation of demons after his virtue has become very clearly known through God's admirable clemency. It is characteristic of an imprudent person to rejoice beyond measure at that time and to vainly and ostentatiously show off the treasures of his good qualities. And thus, one deserves to lose everything that one should have guarded or hidden away in humility so as not to squander it. Let us hear from our teacher, the Lord Jesus Christ, how he has instructed us on the spiritual covering for the eyes, by his words as well as by his example, for he says in Matthew 6, "But when you give alms, do not let your left hand know what the right hand is doing, so that your alms will be given in secret,"[58] as if he were saying, "Do not let ostentation or the love of praise be mingled with your good deeds." And in the same passage: "When you pray, go into your room and pray to your father in secret with the door shut."[59] And this he not only taught us with

52 Gregory the Great, *Homiliae in Evangelia*, 1.11.1, ed. Raymond Étaix, CCSL 141 (Turnhout, 1999), 141.

53 *Vit. pat.*, 3.114, PL 73:781D; ibid., 5.8.19, PL 73:909A.

54 Ibid., 3.113, PL 73:781C; ibid., 5.8.20, PL 73:909A.

55 Cf. Exod 4:6–8.

56 Cf. Isa 38:4–9.

57 Cf. Isa 39:1–7.

58 Matt 6:3–4.

59 Matt 6:6.

words, but confirmed it by his example, for in Matthew 17 he taught those disciples before whom he was transfigured, "Tell no one about the vision"[60] you have seen, and so on. There is a mystical lesson in this passage, as well, for after the triumph he had achieved over the devil during the mystery of his passion, he wanted to be concealed in a tomb for the sacred space of three days, so that after you have triumphed over an enemy you would conceal this fact in the hidden depth of your conscience. As the apostle Paul also says in Colossians 3: "You have died, and your life lies concealed with Christ in God."[61]

This, too, should be noted concerning the eyelids, as Aristotle says in book 4 of *On Animals*, namely, that birds close their eyes with their lower eyelids, but animals that move on their feet with their upper eyelids.[62] Saintly and devout people are signified by the birds; by the animals that move on their feet, worldly people who are wise about matters on the earth. Thus, the eye of animal-like people is closed to heavenly matters, but opened to worldly matters by desire, as in the Psalm: "They have set their eyes to turn to the ground."[63] And concerning the execrable old men it is said in Daniel 13: "They turned their eyes to the ground so they would not see heaven."[64] Spiritual people, on the other hand, do not look at worldly matters with love, but with contempt; they give their attention very enthusiastically to the contemplation of heavenly matters, speaking with the apostle Paul, in Philippians 3: "Our dwelling-place is in heaven."[65] And for this reason the blessed Martin, "with eyes and hands always raised to heaven, said, 'Grant it to me to see heaven!'"[66] And again, one sings about him: "No longer does he deem it worthy to see the earth; his whole being gazes longingly at heaven."[67]

60 Matt 17:9.

61 Col 3:3.

62 Cf. Avicenna, *De animalibus*, 12.11, in *Opera*, 49va.

63 Ps 16:11.

64 Dan 13:9.

65 Phil 3:20.

66 Sulpicius Severus, *Epistolae*, 3.14–15, ed. Karl Halm, in *Libri quï supersunt*, CSEL 1 (Vienna, 1866), 149; cf. Odo of Cluny, *Matins Antiphons for the Feast of Saint Martin*: *In Primo Nocturno*, 6, in *Historia Sancti Martini*, ed. Martha Fickett, Wissenschaftliche Abhandlungen / Musicological Studies, 65/11 (Ottawa, 2006), 7 (PL 133:513B). The reference is to Saint Martin of Tours (ca. 316–ca. 397).

67 Odo of Cluny, *Matins Antiphons for Saint Martin*: *In Prim. Noct.*, 5, ed. Fickett, 6–7 (PL 133:513A).

The sixth moral lesson. That the eyes, that is of the intellect and of the affections, should be turned to God.*

Sixth, the eye is located in the head. It is said in Ecclesiastes 2: "The wise man has eyes in his head."[68] Our head is Christ, from whom sensation and motion flow into us spiritually. Thus, we ought to locate our eyes in this head, namely the eye of the intellect through cognition and the eye of the affections through love, since wherever your love is, there is your eye.[69] For this reason the prophet David located his eyes in the head which is Christ, for David said, "My eyes are always turned to the Lord."[70] Our intellect ought to be located in Christ so that our life will be able to conform to the life of Christ, whom we have examined, and to his way of life, just as the author correctly writes in 2 Chronicles 20*, based on the fact that he directs his eyes to the divine exemplar: "Since we do not know what we ought to do, we who remain possess only this: that we turn our eyes to you."[71] Our affections should also be located in Christ, as our head, so that by paying attention to what he did for us, what he accomplished, and how many favors he conferred on us, we will love him above everything else.

But many people are like a pig that devours the fruit of a tree and never raises its eyes to the tree. It is said in the Psalm: "Just as the eyes of a maid are on the hands of her mistress,"[72] and so on. For plenty of people are like paupers who look at the hands of an almsgiver because of his gifts, and once they have received alms they turn their backs on the almsgiver and forget about him from that point on. So, too, many who have received favors from the Lord turn away from him with ingratitude. And the verse in Ecclesiasticus 27* can be said about such people: "Whoever seeks to become rich turns his eye away."[73] For after they have become rich, they turn the eye of their consideration away from God and they abandon him in their forgetfulness, Deuteronomy 32: "He grew fat, well loved, and he became disobedient. He forsook God, his maker, and abandoned God, his salvation."[74]

68 Qoh 2:14.

69 Cf. Matt 6:21, Luke 12:34. Peter expresses the idea that "wherever your love is, there is your eye" on three occasions: see also below, chapt. 7.11, p. 71, and chapt. 8.5, p. 92.

70 Ps 24:15.

71 2Chr 20:12. *Conformari* is used here as a deponent verb = "to conform to."

72 Ps 122:2. The verse continues: "... so our eyes are on the Lord our God."

73 Sir 27:1.

74 Deut 32:15.

The ingratitude of such people will be confounded if they give thought to the gratitude of irrational animals. The commentator on the *Book of Animals* relates that a man who was the head of a family had at one time domesticated a viper to such a degree that every day it came out from its pit to the man's table at breakfast time begging to be provided with food, and after the meal returned to the pit from which it had come. After some time had gone by, it bore two offspring which it brought to its host's table to be fed. One of these offspring, however, acted as if it were untamed and bit and poisoned the son of the head of the family. The mother viper perceived this and as if punishing the ungrateful offspring killed it in the sight of all and then, as if capable of reason, withdrew with the other one in confusion and shame.[75]

Now, to completely confound human ingratitude, I will relate yet a second *exemplum*. This is written in the *Deeds of the Romans* and likewise the *Policraticus** relates it in book 5: Once in the city of Rome a certain man named Andronicus, condemned to death because of his misdeeds, was thrown to a lion in order to be devoured. As soon as the lion saw him from a distance, it unexpectedly stood still as if in astonishment and then walked towards him peacefully, wagging its tail gently in the manner of fawning dogs as if full of joy and, standing very close to the body of the man, who was by now nearly a corpse for fear, it gently licked his legs and hands with its tongue. When the man became aware of the lion's blandishments, little by little he opened his eyes just a crack to look at the lion and he began to recover the courage he had lost through fear. After this mutual recognition, as it were, had taken place, Caesar observed that the man and the lion were happy and rejoiced together, while the people who had come to the spectacle applauded. Then he called the man to him and questioned him closely about why he alone had been spared by this most ferocious lion that normally devoured on the spot every other person thrown to it. And the man related a miracle, saying: "When my master received the province of Africa to rule as proconsul, I fled to an uninhabited place in the desert because of his cruelty and beatings and because I was unable to endure his rule, and I took shelter in a well-hidden cave I found there. Not long afterwards, the lion came to the cave

75 Cf. Vincent of Beauvais, *Speculum naturale*, 20.21, in *Speculum quadruplex; sive Speculum maius* (Douai, 1624; reprint Graz, 1964), vol. 1, 1472e (cf. Pliny the Younger, *Naturalis historiae libri xxxvii*, 10.74[96].208, ed. Ludwig von Jan and Karl Mayhoff [Leipzig, 1865–1898; reprint 1967], vol. 2, 282). The narrative is also found as an addition to Odo of Cheriton's *Fabulae* in London, British Library MS. Harley 219 (s. xv). See *Odonis de Ceritona fabulis addita*, 1.4, ed. Léopold Hervieux, in *Les fabulistes latins depuis le siècle d'Auguste jusqu'à la fin du moyen âge*, 2nd ed. (Paris, 1893–1899; reprint New York, [1965]), vol. 4, 363–64.

with a wounded and blood-stained paw, roaring terribly and moaning with the pain of its wound. I was terrified beyond measure by this sight and frightened in nearly my entire being. But after the lion saw me when it entered its dwelling-place, it approached gently and tamely, displayed its wounded paw, and held it out to me in order to be cured. Then I carefully pulled out the enormous piece of wood sticking from the sole of its paw, stemmed the watery pus by putting a compress on the wound, wiped the blood away, and gently dried its injury. When it felt relief because of my medical attention, it put its paw in my hands, lay back, and was quiet. And from that day on for three years the lion and I remained in that den and ate the same food. When it caught wild animals while hunting, it used to offer me the richest parts, which I ate after cooking them in the noon-day sun, since I wasn't able to make a fire. But after I grew weary of this way of life, I left the den while the lion was out hunting, and after being chased for three days, I was captured by soldiers, was led back to the master from whom I had formerly fled, and was brought from Africa to Rome where my master saw to it that I was condemned to death as a capital offender and thrown to the wild beasts. So this is the lion I formerly had as a friend, that – as you see – shows kindness and repays me for my faithful friendship and for the medical attention I gave it." Then Andronicus was released from punishment on Caesar's orders and given this same lion as a companion. From that day on, when the lion and the man wandered through and around the city as firm friends, the people would say, "This is the lion that was a man's host; this is the man who was a lion's physician."[76] Thus, people who are unmindful of divine favors should be ashamed when they hear of this kind of gratitude in wild animals towards those who are beneficent to them.

The seventh instruction. That one should visualize Christ's wounds in one's memory.*

Seventh, the eye is receptive to the form or *species* of the visible object, for which the following is one manifest proof among others: It is proven at the beginning of the *Perspectiva* that if in broad daylight someone looks steadily at the sky from a location inside a house that has a large opening above it exposed to the heavens and then returns to a darkened place or closes his eyes, he will see an image of that opening with its shape.[77] The

76 The story of Andronicus and the lion is taken from John of Salisbury, *Policr.*, 5.17, ed. Webb, vol. 1, 361–63; cf. Tubach, *Index exemplorum*, no. 215.

77 Cf. Alhacen, *De aspect.*, 1.1 [4.4], ed. and trans. Smith, in *Alhacen's Theory of Visual Perception*, vol. 1, 4 (Latin text), vol. 2, 343 (English translation).

opening which all of us should pay attention to and look at frequently is Christ's side pierced on the cross. For this reason Revelation 1 reads: "Every eye will see him; they, too, who pierced him,"[78] and Zechariah 12: "They will look at him, whom they have pierced."[79] He said "they have pierced" in the plural, for all of us have pierced Christ, or rather we have crucified him, since Christ was crucified for everyone. Let each and every person enter the house of his conscience and consider Christ's wounds with the eyes of his mind, so that in his own small measure he might conform himself to the suffering Christ. It is written in the *Book on Mixing Elements* that if a murderer looks at the man he has killed, or if he appears in the man's presence, blood will immediately flow out of the murdered man's wounds.[80] If this is true, it can be inferred that if we killed Christ (since he was killed for us), then whoever is not affected by his blood as if it had recently flowed from his wounds is not looking correctly at him. But few look at this opening, and for this reason the Lord laments in Proverbs 1, "I stretched out my hand" – add here: pierced on the cross – "and there was no one who paid any heed."[81]

Oh, how devoutly did the blessed Bernard look at these openings! For this reason, when he expounds the verse in Song of Songs 2: I came to "openings in the rock, to caves in the rock face"[82] in *Sermon 61 on the Song of Songs*, he says as follows: "I dwell there in security since 'he sets my feet on the rock.'[83] The world rages, the body is oppressive, the devil lays snares. I do not fall, 'for my foundation is on a firm rock.'[84] I have committed a great sin; my conscience will be disturbed but not thrown into turmoil, since I will remember the Lord's wounds. Indeed, 'he was wounded for our transgressions.'[85] But for my part, whatever I lack in myself I obtain for myself in faith from the Lord's wounds since they flow with mercy, nor is there a lack of openings through which they can flow out. For 'they stabbed his hands and feet,'[86] and they pierced his side with a spear, and through these slits I am able 'to suck out for myself

78 Rev 1:7.

79 Zech 12:10.

80 Cf. Urso Salernitanus, *De commixtionibus elementorum libellus*, 6.10, ed. Wolfgang Stürner, Stuttgarter Beiträge zur Geschichte und Politik, 7 (Stuttgart, 1976), 199 (cf. Tubach, *Index exemplorum*, no. 5395).

81 Prov 1:24.

82 Cant 2:14.

83 Ps 39:3.

84 Luke 6:48.

85 Isa 53:5.

86 Ps 21:17.

honey from the rock and oil from the hard stone,'[87] i.e., 'to taste and to see how sweet the Lord is.'[88] For the nail (*clavus*) which pierced him has become a key (*clavis*) for me to open the lock so that I can see the Lord's will. The secret compartment of his heart is revealed through the openings of his body, laid bare are 'the viscera of our God's mercy, in which the morning sun has visited us from on high.'[89] Thus, I will go to those storerooms crammed to the brim, and at* the admonition of the prophet I will leave the cities and dwell on the rock, and I will be 'like a dove nesting in the highest opening of the cave.'"[90] This passage was taken from Jeremiah 48.[91] And in the *Lenten Sermon on "A thousand may fall at your side,"*[92] the same Bernard says: "Oh, that I might be worthy to be like the dove that dwells 'in the openings in the rock'[93] and in the cleft in Christ's side!"[94]

The eighth moral instruction. Against fools neglecting their own deeds and rashly judging others' deeds.

Eighth, the corporeal eye is not able to examine itself, but it can examine another object, and in this the eye of the flesh differs from the eye of the mind, for the body's eyes cannot see themselves, but they see other things. On the other hand, the eyes of the heart are able to judge themselves infallibly, but not others. For this reason, people who judge others and not themselves turn their mental eyes into corporeal ones, concerning which, Job 10: "Do you have eyes of flesh?"[95]

And therefore those people are foolish who rashly presume to judge others' deeds and thoughts that they are unable to perceive. But today one finds quite a few like them. For this reason, Augustine says in book 9 of the *Confessions*: "It is human nature to be overly curious to know someone else's life, but slothful in knowing one's own."[96] And the blessed Bernard in the *Meditations*: "Many know a great deal and do not know

87 Deut 32:13.

88 Ps 33:9.

89 Luke 1:78.

90 Bernard of Clairvaux, *Serm. sup. Cant.*, 61.2.3–4, 61.2.6, ed. Leclercq, Talbot, and Rochais, in *Opera*, vol. 2, 150–52.

91 Jer 48:28.

92 Ps 90:7.

93 Cant 2:14.

94 Bernard of Clairvaux, *Sermones in Psalmum "Qui habitat,"* 7.15, ed. Jean Leclercq and Henri M. Rochais, in *Opera* (Rome, 1966), vol. 4, 424.

95 Job 10:4.

96 Augustine of Hippo, *Conf.*, 10.3.3, ed. Verheijen, CCSL 27:156.

themselves; they inspect others and forsake themselves."[97] And "Gregory Nazianzen: 'Nothing seems to be more delightful to people than to talk about the affairs of others or to pay attention to others' affairs, but this appears to be inconsistent with the judgment of a tranquil mind.'"[98] But as Jerome asserts to Asella in *Letter 83*: "It is dangerous to make judgments about another person's servant and there is no easy forgiveness for saying vicious things about those who are just."[99] These people are like the foolish student who possesses a book written in the Latin alphabet, which he well understands and from the study of which he can greatly profit, and nevertheless he never wants to look into it, but rather wastes his time daily on a book in Greek or Hebrew, which is utterly unknown to him. Such people appear to strip off their human shape and put on the shape of a lynx whenever they say that they perceive the secrets of others' hearts.[100] Nevertheless, as it is said in 1 Corinthians 2: "For among men, who knows what a man is if not the spirit of that man within him?"[101]

And they are truly foolish, since as Cicero says in the *Tusculan Orations*: "It is a characteristic of foolishness to perceive others' vices and to forget one's own."[102] These are the people to whom the Lord says in Luke 6, "Why*, then, do you see a speck in your brother's eye, but you take no notice of the beam in your own eye? Or how could you say to your brother, 'Brother, let me take the speck out of your eye,' not seeing the beam in your own eye?"[103] For these people do not see grave sins in themselves, but in others they see matters which are insignificant or perhaps not even there at all. For this reason, the poet Horace says:

> Since you investigate your own evils blinded by eyes smeared with ointment,
> Why do you perceive your friends' vices so exactly?[104]

97 Ps.-Bernard of Clairvaux, *Meditationes piissimae de cognitione humanae conditionis*, 1.1, PL 184:485A.

98 John of Salisbury, *Policr.*, 7.prol, ed. Webb, vol. 2, 91–92.

99 Jerome, *Ep.*, 45.1.2, ed. Hilberg, CSEL 54:324.

100 Roger Bacon, *Persp.*, 1.9.2, ed. and trans. Lindberg, 130–32, asserts that Boethius attributes to Aristotle the claim that a lynx can see through a wall, though Boethius' reference is actually to Lynceus, one of the Argonauts, who was endowed with the gift of especially acute sight. See Boethius, *Phil. cons.*, 3.pr.8.10, ed. Bieler, CCSL 94:48.

101 1Cor 2:11.

102 Cicero, *Tusculanae Disputationes*, 3.30.73, ed. Max Pohlenz (Stuttgart, 1918; reprint 1967), 354.

103 Luke 6:41–42.

104 Horace, *Sermones*, 1.3.25–26, ed. Edward C. Wickham, 2nd ed., ed. Heathcote W. Garrod, in *Opera* (Oxford, 1901) (cf. Walther, *Proverbia*, no. 4553).

Such people are similar to thieves who neglect open coffers, that is to say, their own conscience, which is accessible and familiar to them, and instead they look for locked coffers, that is to say, other people's consciences, which they cannot see. Seneca: Thieves look for things which are closely guarded; they pass by what is unlocked.[105] Such people can appropriately be compared to slumberers. For someone who falls asleep closes his eyes to himself when he lays his head down, but he opens his eyes to see others when he raises his head up. But let these people hear what is said about them in Proverbs 23: "Drowsiness will be clothed in rags."[106] On account of this sleep, by which they have neglected themselves and want to judge others, they will be dressed in rags, i.e., in the clothing of damnation, and those who attempt to confuse others by rashly judging them will be clothed, as it were, "in their own confusion as in a cloak."[107]

An eye located in the light does not see someone in darkness, but rather the other way around. Thus, sinners persisting in darkness recklessly presume to judge those living in the light of grace, but the just, illuminated* by the light of heaven, do not know how to judge sinners. And therefore, in order for us to be children of light, we should not judge sinners as soon as we see them, but rather let us weep and suffer with them in their miseries, since either we have fallen ourselves in similar circumstances, or we could fall. In the *Life of Saint John the Almoner* one reads that among other good qualities the saint also had the characteristic of judging no one, but rather when he heard people making judgments about others he would correct them, saying, "You are sinning first of all against the command of the Lord who says, 'Do not judge so that you will not be judged';[108] and second because you do not know what state the person is in now whom you have come to know as a sinner."[109] For we have often witnessed the sin of a thief or a perjurer or a fornicator, but we have not seen his penance which has been accepted by God. One also reads in the *Lives of the Fathers* that the holy hermit Moses, who had been invited to Scythia* to judge a criminal brother there, carried a very old basket filled with sand. When he was asked what this was, he answered, "They are my sins following me, and I don't see them, and today I come to judge someone else's sins."[110] This man, who was never

105 Cf. Seneca, *Ep. mor.*, 68.4, ed. Reynolds, vol. 1, 198.

106 Prov 23:21.

107 Ps 108:29.

108 Luke 6:37.

109 Leontius of Naples, *Vita s. Ioannis Eleemosynarii*, 46, trans. Anastasius Bibliothecarius, PL 73:377A–B.

110 *Vit. pat.*, 5.9.4, PL 73:910B; cf. ibid., 3.136, PL 73:786C–D (cf. Tubach, *Index exemplorum*, no. 4413).

so bold as to rashly judge a sinner living in darkness, had rightly been given a place in the light of grace. Anselm says in the *Similitudes*: "Paying attention to someone else's sin brings about many evils in various ways. For if the one interested in another's faults is a sinner himself, he will persist in his sin on the other person's example, but if he is doing penance for sin, his penance will grow cold from the same example, since whether he was tempted by the same sin or by a different one, he will be overcome more easily because of that example. Even if he was habitually free of temptation, he will be tempted by the very fact that he is concentrating on that person's vice, and if he is an upright person, he will be exalted in pride by considering himself better than that person. Even if he had loved him earlier in God, he will now begin to hate him. Another person's sins are not something to consider lightly, because concentrating on this matter brings about so many evils."[111]

Thus, holy men do not judge others rashly, but know how to judge themselves carefully, being aware that it has been written in Ecclesiasticus 17*: "He placed their eye above their heart."[112] In these words the wise man clearly demonstrates the property of the interior eye that was just mentioned, so that when the eye is said to be placed above the heart, the wise man demonstrates that a person is capable of seeing himself, but he cannot see the secrets in the hearts of others. For only the person whose secrets they are sees those matters and no one else does, except God whose eye perceives everything. For this reason, it is said in Job 28: "The bird did not know the path, nor did he look upon the eyes of the vulture."[113] The sinner is designated by the vulture, "for the vulture gladly walks on its feet, and for this reason some also call it the 'foot slogger' (*gradipes*), since the sinner loves worldly things and gazes longingly at them. Sometimes* the vulture also flies up in the air, as at times a sinner raises his mind to heavenly matters. But no one else knows with what intention he does this, for who gazes at the eyes of the vulture, i.e., at the intention of someone else's thought? The Almighty has reserved this for himself, for he alone knows human beings' thoughts."[114]

111 Ps.-Anselm of Canterbury, *Lib. de simil.*, 111, PL 159:669C–670A (see also Ps.-Anselm of Canterbury, *Liber de humanis moribus per similitudines*, 111, ed. Richard W. Southern and Franciscus S. Schmitt, in *Memorials of St. Anselm*, Auctores Britannici Medii Aevi, 1 [Oxford, 1969], 82).

112 Sir 17:7.

113 Job 28:7.

114 Hugh of Fouilloy (Ps.-Hugh of Saint-Victor), *Aviarium*, 2.43, ed. and trans. Willene B. Clark, in *The Medieval Book of Birds. Hugh of Fouilloy's Aviarium*, Medieval & Renaissance Texts & Studies, 80 (Binghamton, NY, 1992), 202–03.

*The ninth moral instruction. That sinners look down on heavenly matters and esteem worldly matters**.

Ninth, the eye has knowledge of visible things in different ways according to variations in their distance, for if it sees an object far away, it judges it to be small, but if up close, it judges it to be large. For this reason, we judge the sun to be moderate-sized, which we perceive from afar, although it may be demonstrated by the most exact proof to be one hundred sixty-six times as large as the earth.[115]

In similar fashion, sinners are removed very far from God and a heavenly reward by the distance of sin, for salvation is far away from sinners. They look down on God and a heavenly reward, while they esteem worldly goods, which are close to them through love. For this reason, it is said in Numbers 24: "Balaam son of Beor spoke, the man whose eye was closed up,"[116] "who falls in a trance with open eyes":[117] "I will see him, but not now; I will look upon him, but not close up."[118] Since indeed sinners fall and have descended to temporal matters by their love for them (for their eyes are wide open to worldly things, but closed to heavenly ones), they look on God as if from afar, and therefore they despise him, not judging him according to his greatness. For this reason Pharaoh, too, distanced from God by sin, said, "Who is the Lord, that I should hear his voice? I do not know the Lord and I will not let Israel go."[119] For this reason, the Lord says to such people in Isaiah 33 as if they were a long way off, "Hear, you who are far away, what I have done, and know my strength, you who are near."[120] For a similar reason, these people consider heavenly matters to be trifling: Due to a defect of love, they contemplate them as

115 Roger Bacon quotes the astronomer Thabit ibn Qurra as making this calculation of the size of the sun in *Opus majus*, 4.4, ed. John Henry Bridges, in *The 'Opus majus' of Roger Bacon* (Oxford, 1897), vol. 1, 234 (in the section on the Application of Math to Sacred Subjects): "Nam dicit [Thebit], quod Sol est centies sexagesies sexies aequalis Terrae" (For Thabit says that the sun is equal to the earth 166 times). Cf. Ptolemy, *Almagestum*, 5.16, in *Almagestum seu magnae constructionis mathematicae opus plane divinum Latina donatum lingua ab Georgio Trapezuntio ... per Lucam Gauricum ... recognitum* (Venice, 1528), 51v, the first Latin translation of the Greek version of the *Almagest*, where the sun is measured to be 170 times the size of the earth (cf. *Ptolemy's Almagest*, 5.16, trans. Gerald J. Toomer [London, 1984], 257).

116 Num 24:3.

117 Num 24:16.

118 Num 24:17.

119 Exod 5:2.

120 Isa 33:13.

if far removed from themselves. In the guise of such people it is said in Isaiah 38: "My eyes are weakened looking upward."[121]

However, they judge worldly matters to be sizeable, insofar as they are very close to their desire. And for this reason, in Jeremiah 22 the Lord says of the avaricious person, "He says, 'I will build for myself a large home and a spacious upper dining room.'"[122] And these words follow: "For your eyes and your heart are set on avarice."[123] Namely, to the eyes of the avaricious person, the worldly home that he sees up close seems to be large, but he considers the heavenly home that he looks at from far away to be small. Nevertheless, as it is written in Baruch 3: "Oh Israel, how great is the house of God and how vast a size is the place of his possession! It is large and has no end, lofty and immense."[124] On the other hand, heavenly things appear large to the eyes of the holy, as if located not far off, and therefore because God was very close to his desire, David knew the greatness of God. For that reason, he also said, "For I know that the Lord is great, and our God is above all gods."[125] And concerning the copiousness of heavenly reward, the same author said, "How great is the multitude of your sweetness, Lord!"[126] But since worldly matters are far away from the hearts and desire of those who are holy, they seem small to them, or even as if they were nothing at all, Jeremiah 4: "I looked at the earth, and behold, it was void and worthless,"[127] and so on.

For a similar reason the proud think of themselves as grand, but the humble consider themselves small. Once, this question was raised among scholars: How is it that the more perfect and holy someone is, the more humble he is; and the greater a sinner, the prouder he is – since it appears it ought to be just the opposite? For the holy person seems to have more cause by which he ought or is able to be elevated, but the sinner more cause to be humble, since he possesses no good qualities. One person spoke up promptly to say that it ought to be just the other way around: "For humility brings a person closer to God, but pride distances him from God." We see in the structure of the universe that the smaller a circumference is, the closer it is to the center point. But Dionysius says in the *Celestial Hierarchy* that God is like the center point, and

121 Isa 38:14.
122 Jer 22:14.
123 Jer 22:17.
124 Bar 3:24–25.
125 Ps 134:5.
126 Ps 30:20.
127 Jer 4:23.

intelligible spirits move around the center in the manner of a circumference.[128] It works in the same way for human beings who in their own way move around God, according to the verse in Acts 17: "In him we live, we move, we exist."[129] Thus, the narrower someone's circumference, i.e., the more humble he is, the closer he is to God, and the wider someone's circumference is through pride, the farther away he is from God. The Psalm: "He regards the humble," namely from close up, "and he acknowledges the lofty" (i.e., the proud) "from a long way off."[130] And in Ecclesiasticus 15: "[Wisdom] is far distant from pride."[131] Now, it is true that the more distant someone is when seen, the smaller the person appears to be. For if someone looks at a tower from a distance, the farther he is from it, the smaller it appears to be, and the nearer he is, the larger it appears to be. And when he comes to the base of the tower, he will judge it to be its largest. Holy people are located close to God, and for this reason one of them requested, "Set me free, Lord, and place me next to you,"[132] or rather they sit at the feet of the Lord as his students, Deuteronomy 33: "Those who draw near to his feet receive his instruction."[133] And on that account, when they see themselves as dwarfs standing next to giants, they consider God great and themselves small. But sinners, being distant from God through sin, magnify and exalt themselves when they pay no attention to the greatness of God.

The tenth moral instruction. That we should freely share with others the goods that we possess.*

Tenth, the eye establishes the direction for the entire body. For this reason, the eyes are also called lamps, since they receive light externally and then give it back by sharing what they have obtained. For this reason when he defines the eye, Secundus the philosopher says: "The eye is the leader of

128 Cf. Ps.-Dionysius the Areopagite, *De caelesti hierarchia*, 15, trans. Robert Grosseteste et al., ed. Phillipe Chevallier et al., in *Dionysiaca. Recueil donnant l'ensemble des traductions latines des ouvrages attribués au Denys de l'Aréopage* … ([Bruges], 1950), vol. 2, 1034–35 (cf. Ps.-Dionysius the Areopagite, *The Celestial Hierarchy*, 15.9, trans. Colm Luibheid, in *The Complete Works* [New York and Mahwah, NJ, 1987], 190).

129 Acts 17:28.

130 Ps 137:6.

131 Sir 15:7.

132 Job 17:3.

133 Deut 33:3.

the body, a vessel of light, and the window of the soul."[134] For the eye does not see for itself alone, but for the entire body, nature instructing us in this way to be eager to share liberally with others the good qualities that we have. For this reason, too, when he speaks of wisdom in Wisdom 7, the wise man says: "What I have learned without guile, I also share without envy, and I do not conceal the virtue of wisdom."[135] We are also warned about this in 1 Peter 4: "Since each one has received a gift, make use of it for each other, like good stewards of God's varied grace."[136]

For someone does not possess fewer spiritual goods by sharing them with others. And indeed, Cassiodorus says in his letters, *Epistle 14*: "It should not be difficult to share kindness, because it is not diminished through largess."[137] And Augustine says in book 15 of *The City of God*: "The possession of goodness is not diminished by sharing it with a temporary or permanent partner, since the more broad-based it is, the more harmonious is the joy possessed by the brotherly love of companions. In short, whoever is unwilling to share this possession with others will not possess it at all, and he will discover that the more he is able to love his partner by sharing with him, the more abundant is his goodness."[138] And so the *Anticlaudianus* says in book one:

> [This proposal] goes forth
> For the good of all; it will have a better fate in the light.
> For the good that lies hidden in a dimmer light
> Grows weak, but it will increase all the more if it can enjoy the light.
> Thus, the flower will develop into fruit, the brook will become an overflowing
> River, and the ear of corn will ripen fat for the harvest.[139]

134 *Vita Secundi philosophi*, 33 (trans. Willelmus medicus), ed. Walther Suchier, in *Altercatio Hadriani Augusti et Epicteti Philosophi*, ed. Lloyd William Daly and Walther Suchier, Illinois Studies in Language and Literature, 24/1–2 (Urbana, IL, 1939); reprint in Ben Edwin Perry, *Secundus the Silent Philosopher*, Philological Monographs, 22 (Ithaca, NY, 1964), 98. Cf. Vincent of Beauvais, *Speculum historiale*, 10.71, in *Speculum quadruplex; sive Speculum maius* (Douai, 1624; reprint Graz, 1965), vol. 4, 393a. See below, chapt. 7.11, p. 71. On the medieval Latin tradition of the *Vita Secundi philosophi*, see Perry, *Secundus the Silent Philosopher*, 23–52.

135 Wis 7:13.

136 1Pet 4:10.

137 Magnus Aurelius Cassiodorus, *Variarum libri xii*, 1.14.1, ed. Åke J. Fridh, CCSL 96 (Turnhout, 1973), 24.

138 Augustine of Hippo, *De civitate Dei*, 15.5, ed. Bernhard Dombart and Alfons Kalb, CCSL 48 (Turnhout, 1955), 458.

139 Alan of Lille, *Anticlaudianus*, 1.345–50, ed. Robert Bossuat, Textes philosophiques du moyen âge, 1 (Paris, 1955), 67.

The same author says in book 7:

> Heavenly favor, the treasure chest and miracle of the soul,
> Wants generous owners and disdains misers.
> Locked away, it perishes; offered to others, it is returned; unless it is made a public good,
> It fades away, and it acquires great power as it leaves.[140]

A wise man says: People will see that you commit the fault of ingratitude when you are slow to share with others what everyone well knows you received for free. Indeed, what is worse than to lock this gift up under watchful guard through your negligence? For if it had been given to others, it would have brought an increase in well-being to the giver and the receiver.

Whoever appears to boast that what he has received is his private property and shares it with no one is considered to have robbed his neighbor. It is a blind prudence that does not want to help others when it can. What Bernard says harmonizes well with this idea: "In effect, you retain what belongs to your neighbor if you are someone adorned with the gifts of knowledge and eloquence, but you bind up in a useless or even damnable silence a well-spoken word that could benefit many others, since 'He is cursed who hides away grain from the people,'" Proverbs 11.[141] Now spiritually, just as the corporeal eye shares its light with the rest of the body's members, since it sees not only for itself, but for the entire body, so someone who has been illuminated by the gift of knowledge ought to share it liberally with everyone, so that he can say the verse in Ecclesiasticus 24: "See that I have labored not only for myself, but for all those seeking the truth."[142] In the same way David entreated the Lord, saying, "Light up my eyes."[143] And elsewhere he says, "I will teach the wicked your ways."[144] And Paul, who had lost his eyesight, began to go into the synagogue all the time to preach to the Jews after he received his vision again from the Lord through Ananias.[145] For this reason, Seneca has said in one of his letters to Lucilius*: "I am anxious to pass on everything to you, and I am happy to continue learning so that I may

140 Ibid., 7.237–41, p. 164. The first verse of this excerpt in the printed text of Peter's work differs from what is found in Bossuat's edition.

141 Prov 11:26; Bernard of Clairvaux, *Serm. sup. Cant.*, 18.1.2, ed. Leclercq, Talbot, and Rochais, in *Opera*, vol. 1, 104.

142 Sir 24:47.

143 Ps 12:4.

144 Ps 50:15.

145 Cf. Acts 9:12–20.

teach, nor does anything please me, no matter how excellent or beneficial it may be, if I am forced to know it for myself alone. If wisdom were given to me with the stipulation that I keep it to myself, I would not scorn it, I would abandon it. In fact, there is no pleasure in possessing something good without a companion*."[146]

And just as we see that the corporeal eye does not see less just because it sees for the entire body, but rather that the activity of seeing renders it even more powerful at seeing (for which reason some people go blind from a long stay in prison), so, too, knowledge is not diminished by being shared, but rather it is increased. For this reason Isidore says: "Spread by word of mouth the knowledge you receive with your ear."[147] For when it is passed on to others, wisdom grows more abundant; when it is hoarded, it is diminished. And in the first *Letter of Alexander to Dindimus**, *King of the Brahmins*, one reads this: "When it is communicated, knowledge is a source of abundance to which the suffering of loss is unknown, since knowledge is passed on as a bequest when it is shared with someone else, just as when you kindle many lamps or firewood from one torch you will not diminish its original substance. Indeed, it has the capability of shining more brightly whenever it finds opportunities to offer more to others."[148] But sometimes the Lord withdraws the gift of knowledge from certain people when they do not share the knowledge they have received with others, but act ungenerously and hide it away. For this reason, Bernard refers to himself in *Sermon 82* on the Song of Songs*: "When the Spirit was prompting him on what he should say and he held a few of these ideas in reserve so that he would have something to say about one of them when he discussed the subject again (and yet his heart was not unfaithful, though it was not confident enough), behold, a voice came to him saying, 'As long as you hold onto this idea, you will not receive another one.'"[149] Therefore, like the corporeal eye, let us share our good qualities with others, for Cicero says in *On Duties*: "One thing should be everyone's object: to make the good of one human being identical with that of each and every other one."[150]

146 Seneca, *Ep. mor.*, 6.4, ed. Reynolds, vol. 1, 10 (with variations from Seneca's text).

147 Isidore of Seville, *Synonyma*, 2.67, ed. Jacques Elfassi, CCSL 111B (Turnhout, 2009), 118 (PL 83:860D–61A).

148 Cf. *Epistola Alexandri regis ad Dindimum regem*, ed. Friedrich Pfister, in *Kleine Texte zum Alexanderroman*, Sammlung vulgärlateinischer Texte, 4 (Heidelberg, 1910), 10.

149 Bernard of Clairvaux, *Serm. sup. Cant.*, 82.1.1, ed. Leclercq, Talbot, and Rochais, in *Opera*, vol. 2, 292.

150 Cicero, *De officiis*, 3.6.26, ed. Michael Winterbottom (Oxford, 1994), 119. The end of the sentence in Peter's text varies from what is found in the standard text of Cicero's work.

*The eleventh moral instruction. That the state of someone's mind is signaled through his body's eyes**.

Eleventh, the eye is indicative of the secrets of the mind. For this reason, the movement of the eye is also a sign of the interior state of the mind, as is made clear by Aristotle in book 12* of *On Animals*: If the eye moves too much, it is a sign of mental instability; if it is too sluggish, of an obstinacy of the will; if it is moderate in movement, it is a sign of the mind's ease in apprehending and of the required grasp of what it has apprehended.[151] For some are fervent at the beginning, but they stop after having initiated a good deed, since they believe in it for awhile, but as soon as they are tempted, they abandon it; others are lukewarm and languid; while yet others begin a good deed easily and persevere steadfastly in what they have begun. For since "the eye is the window of the soul," as Secundus the philosopher says,[152] we can very often assess the state of someone's mind through his eye. For the exterior eye reveals whatever someone might hide within himself. For this reason, one reads of Ahasuerus in Esther 15: "Then he looked up and with glowing eyes showed the passion in his heart."[153] And Ecclesiasticus 19: "You can tell a person by his look,"[154] and so on. For wherever your love is, there is your eye.[155] And therefore the eye is easily turned to where a person's heart is inclined, as it is said in Job 31: "If my eye has followed my heart."[156]

And therefore it is entirely suitable for an upright person to keep his eye far away from obscene and shameless spectacles, lest the incontinence of his eye be taken to indicate a shameless mind. For this reason, Cicero says in *On Duties*: "When Pericles and* Sophocles both sat together conferring about public issues and a handsome youth happened to walk by, Sophocles said, 'Oh, Pericles, look at the handsome boy.' Pericles responded, 'It behooves a praetor to have not only continent hands and tongue, but also continent eyes.'"[157] For just as we judge the freshness of a fish by its eye (for a fish is not fresh if it has a putrid eye), so someone is also not considered chaste who has an immodest eye. For as Gregory

151 Cf. Aristotle, *De hist. animal.*, 1.10, trans. William of Moerbeke, ed. Beullens and Bossier, 19; idem, *Hist. animal.*, 1.10 (492a10–12), ed. Balme, 71.

152 *Vita Secundi phil.*, 33 (trans. Willelmus medicus), ed. Suchier, 98. See above, chapt. 7.10, p. 68.

153 Esth 15:10.

154 Sir 19:26.

155 Cf. Matt 6:21, Luke 12:34. See also above, chapt. 7.6, p. 57, and below, chapt. 8.5, p. 92.

156 Job 31:7.

157 Cicero, *De off.*, 1.40.144, ed. Winterbottom, 60.

says: An immodest eye is the messenger of an immodest heart.[158] And the versifier says:

> A skittish mind, a wandering* eye, a restless foot:
> These are signs of someone from whom I have no hope of good.[159]

*The twelfth moral instruction. That it is sometimes useful to leave external matters behind and return to internal ones**.

Twelfth, the eye is drowsy after being awake for a long time, and for this reason it is said concerning the apostles in Matthew 26: "He found them sleeping,"[160] for their eyes were heavy. For after the eye of the flesh has been occupied for a long time with the sight of external matters, it is necessary for the animating (*animalis*) spirit, which had devoted itself to these things, to turn inward, and then an animal sleeps and is quiet. In this fact the nature of the eye teaches us that when a person busies himself with taking care of external matters, it is advantageous for him to return to internal ones, and thus he will be able to sleep in a spiritual slumber and will rest delightfully in God. And that is what the Psalmist says: "If I give sleep to my eyes and drowsiness to my eyelids, until I find the place of the Lord."[161] And elsewhere: "At peace in that very place I will sleep and I will rest."[162] Therefore, the Lord warns us in Isaiah 26, "Go, my people, into your bedchambers, and close the door on yourselves,"[163] as if he would say, "Close the door of your external senses and enter the home of your conscience." And at that point our internal being sleeps in a slumber of contemplation, in which the external senses are held in check and kept from their activities so that they can say what the bride says in Song of Songs 5: "I sleep and my heart keeps watch."[164] Namely, spiritual people are said to sleep with their internal* eyes open, like lions and hares. For this reason Paul, who was externally struck blind, was enlightened internally.

Thus, on the example of the corporeal eye a person should be redirected from external matters to internal ones. In a symbolic enactment of this idea, when a deacon is about to read the Gospel, he initially carries

158 Cf. Thomas the Cistercian, *Commentarii in Cantica Canticorum*, 2, PL 206:148D–49A.

159 *Sententia communis* (Walther, *Proverbia*, no. 12215).

160 Matt 26:40.

161 Ps 131:4–5.

162 Ps 4:9.

163 Isa 26:20.

164 Cant 5:2.

the book in his hands, but when he returns [from the lectern], he is empty-handed to indicate that after the duty or activity of preaching, he should devote himself to contemplation.[165] And therefore a prelate, who is considered to act like an eye for the body of the church, should return to himself internally after he has devoted himself externally to the people entrusted to his care, following the example of the cautious shepherd Jacob: After he had attended to the feeding of Laban's sheep for a long time, he said, "It is just that I also look after my own house now and then."[166] For this reason, Bernard wholesomely warned Pope Eugene in *On Consideration*: "What benefit is it to you to gain everyone if you lose yourself? For that reason, when all possess you, you, too, should be one of the possessors."[167]

This activity is necessary not only for a prelate, but also for anyone else, so that from time to time he should leave external matters behind and return to himself internally. For this reason, in his letters Seneca says in *Epistle 15*: "Whatever you do, return quickly from the body to the soul. You must exercise it day and night."[168] And in *Letter 78** he says: "The great and prudent person separates his soul from his body, and spends a great deal of time with the better and long-lasting part."[169] And in the *Meditations* the blessed Bernard says: "Wherever you are, be your own master. Do not surrender yourself to [external] matters, but adapt to them."[170] But there are many people who have an eye that always wanders externally, nor do they ever want to return to their own conscience – as it were, to a house bereft of every spiritual gift. And in this way, they are like the actor who is reluctant to live in his own house and more willingly frequents another's. For just as the hearth in the actor's house is cold, since the actor always stays away from his home and has no container there for

165 It was the duty of the deacon to read the Gospel at Mass; cf. William Durand of Mende, *Rationale divinorum officiorum*, 2.9.6, ed. Anselme Davril and Timothy M. Thibodeau, CCCM 140 (Turnhout, 1995), 159.

166 Gen 30:30.

167 Bernard of Clairvaux, *Tract. de consid.*, 1.5.6, ed. Leclercq and Rochais, in *Opera*, vol. 3, 400. Cf. Matt 16:26.

168 Seneca, *Ep. mor.*, 15.5, ed. Reynolds, vol. 1, 39.

169 Ibid., 78.10, vol. 1, 251.

170 Ps.-Bernard of Clairvaux, *Medit. piissimae*, 6.18, PL 184:497A. The last sentence in the PL text of Ps.-Bernard reads: "Noli te rebus tradere, sed commodare." *Pr1* and *P7P8* agree in using *commendare* as the final verb, but a confusion between the two verbs, and thus an analysis of them as in some way synonymous, is not an unusual occurence. See Egidio Forcellini, *Lexicon totius latinitatis*, rev. ed. Giuseppe Furlanetto, Francesco Corradini, and Giuseppe Perin (Padua, 1864–1898; reprint 1940), vol. 1, 714 (q.v. *commodo*).

fire nor a bowl for washing, so likewise in someone's heart who wanders every day among useless things there is neither the fire of devotion nor a bowl wet with tears of remorse, and for this reason the sinner does not willingly enter his conscience, since he, too, would find nothing there besides his own evil. And therefore he can properly say this verse in Job 7: "And my eye will not return to see good things."[171]

171 Job 7:7.

CHAPTER EIGHT

On Seven Distinctions of the Eyes According to the Distinction of the Seven Capital Vices*[1]

Since they always look at great and high-flown matters, the eyes of the proud are said to be arrogant* and conceited.[2] For haughty eyes display a conceitedness of the spirit, because as Gregory says in the *Morals*: "Pride in the heart is extended externally to the limbs and is shown through the eyes as a sin."[3] And Architrenius, describing the state of the proud man, says:

> Yet, unless all things agree with his will, they make his face
> Grow hot, and his flashing eyes foretell his wrath,
> And by tapping on his nose with his finger, he threatens great deeds.[4]

1 For introductions to the seven deadly sins in the Middle Ages, see *The Seven Deadly Sins: From Communities to Individuals*, ed. Richard Newhauser, Studies in Medieval and Reformation Traditions: History, Culture, Religion, Ideas, 123 (Leiden and Boston, 2007); *In the Garden of Evil: The Vices and Culture in the Middle Ages*, ed. Richard Newhauser, Papers in Mediaeval Studies, 18 (Toronto, 2005); Carla Casagrande and Silvana Vecchio, *I sette vizi capitali. Storia dei peccati nel Medioevo*, Saggi, 832 (Turin, 2000); Richard Newhauser, *The Treatise on Vices and Virtues in Latin and the Vernacular*, Typologie des sources du moyen âge occidental, 68 (Turnhout, 1993); Morton W. Bloomfield, *The Seven Deadly Sins: An Introduction to the History of a Religious Concept, with Special Reference to Medieval English Literature* ([East Lansing], 1952; reprint 1967).

2 Introductions to medieval conceptualizations of pride, and the decisive focus on this sin by Augustine of Hippo and Gregory the Great, can be found in Carole Straw, "Gregory, Cassian, and the Cardinal Vices," in *In the Garden of Evil*, ed. Newhauser, 35–58; John C. Cavadini, "Pride," in *Augustine through the Ages: An Encyclopedia*, ed. Allan Fitzgerald (Grand Rapids, MI, 1999), 679–84; Wolfgang Hempel, *Übermuot diu alte .... Der Superbia-Gedanke und seine Rolle in der deutschen Literatur des Mittelalters*, Studien zur Germanistik, Anglistik und Komparatistik, 1 (Bonn, 1970).

3 Gregory the Great, *Mor. in Iob*, 34.22.46, ed. Adriaen, CCSL 143B:1765.

4 Johannes de Hauvilla, *Architrenius*, 5.5.111–13, ed. Paul Gerhard Schmidt (Munich, 1974), 200. For a full translation, see Johannes de Hauvilla, *Architrenius*,

The prophet David detested those eyes in his fellow human being, saying, "With a proud eye and an insatiable heart, with such I have not eaten."[5] And he also detested them in himself: "Lord, neither has my heart been exalted, nor are my eyes conceited, nor have I walked in great matters."[6] And thus he compares himself to the night owl, saying, "I have become like a night owl in its home."[7] "Now, the night owl does not signify just any righteous man to us, but rather someone who spends his time in the midst of people but hides himself as far as he can from recognition by them, fleeing the light and paying no attention to the glory of human praise."[8] For he has come to know what is written in Isaiah 5: "The eyes of the haughty will be brought down."[9] And therefore David also says, "You will humble the eyes of the proud."[10] Such eyes are extremely hateful to God, who has a mild and humble heart, and for this reason it is said in Proverbs 6: "There are six things which the Lord hates and the seventh his spirit detests: haughty eyes," and so on.[11] Therefore, one reads in Luke 18 that the tax collector was forgiven who did not dare raise his eyes to heaven – as if he were saying, "I am not worthy to see the highest part of heaven because of the multitude of my evils."[12] Nor is it surprising that haughty eyes are displeasing to God, since in this way a man becomes like the devil, about whom it is said in Job 41: "He beholds everything that is haughty and he himself is king over all the sons of pride."[13] And concerning the antichrist it is said in 2 Thessalonians 2: "He raises himself up against every so-called god or object of worship."[14] Thus, in Ecclesiasticus 23 Sirach implored, "Do not give me a haughtiness of my eyes."[15]

---

ed. and trans. Winthrop Wetherbee, Cambridge Medieval Classics, 3 (Cambridge, Engl., 1994); this passage is translated on 123. The translations here are my own. Peter has changed the gender of the passage; Johannes describes Pride as a female personification. Tapping with her finger on her nose is one in a series of bombastic bodily gestures by Pride, but Johannes notes that nothing will come of them. They are empty gestures alone.

5 Ps 100:5.

6 Ps 130:1.

7 Ps 101:7.

8 Hugh of Fouilloy (Ps.-Hugh of Saint-Victor), *Aviarium*, 2.39, ed. and trans. Clark, 174–75.

9 Isa 5:15.

10 Ps 17:27.

11 Prov 6:16–17.

12 Cf. Luke 18:13–14.

13 Job 41:25.

14 2Thess 2:4.

15 Sir 23:5.

There are, however, some who have external eyes that are humble enough, but their internal eyes are excessively haughty: With an immoderate haughtiness they disdain plain facts, while at the same time they also want to examine curiously the hidden matters of God. And for this reason it is just that the dull-witted do not occupy themselves with such matters, and so it is said in Proverbs 23: "Do not raise your eyes to riches that you are unable to possess, since they can make wings for themselves and will fly up to heaven like eagles."[16] That is to say, these matters will conceal themselves from you. For there are many things that are understood less, the more they are scrutinized. Therefore, it is said in Song of Songs 6: "Turn your eyes away from me, since they have made me fly away."[17] On the contrary, such people often lose their internal vision due to their pride, since they become puffed up because of their knowledge. And in this they are similar to bats that see poorly because the crystalline fluid that is necessary for the eye to see is transferred into the substance of the wings. For this reason they are said to have leathery wings.[18] About them Ovid says:

> And yet they hold themselves aloft with transparent wings,[19]

and therefore, because they fly, they have lost their sight, since what is located in the wings has been withdrawn from the eyes. And bats stand for those people who are made conceited by knowledge: The more puffed up they are, the more they are deprived of the light of divine grace. Romans 1 says about them: "In their thinking they have become vain and their foolish hearts are plunged into darkness."[20] And in 2 Kings 6 it is said of the Assyrians, who are interpreted as "haughty ones":[21] "Strike, I pray, this people with blindness."[22]

No wonder if pride blinds the eye of reason, since pride is a spiritual film over the eye, as Gregory explains concerning the verse in Leviticus 21: "If he is someone who has a film over his eye,"[23] and so on: "When the

16 Prov 23:5.

17 Cant 6:4.

18 Cf. Aristotle, *De hist. animal.*, 1.1 and 1.5, trans. William of Moerbeke, ed. Beullens and Bossier, 8 and 14; idem, *Hist. animal.*, 1.1 (487b23) and 1.5 (490a8), ed. Balme, 57 and 64.

19 Ovid, *Metamorphoses*, 4.411, ed. William S. Anderson (Leipzig, 1977), 90.

20 Rom 1:21.

21 Cf. Augustine of Hippo, *De civ. Dei*, 16.17, ed. Dombart and Kalb, CCSL 48:521.

22 2Kgs 6:18.

23 Lev 21:19.

pupil of the eye is black, it sees"; when it is white, it does not.[24] People who claim for themselves the radiance of virtue have a film over their eye, and sometimes they experience the costliness of this loss of sight after carrying out virtuous deeds. As an indication of this fact, Tobit, "coming home exhausted from the burial of the dead,"[25] lost his eyesight, and therefore Gregory says when he speaks about pride: That vice must be greatly feared which is born from victory over the vices.[26] And the proud person is truly blind when he does not understand that he himself is a human being. For the word *homo* ("human being") comes from *humus* ("dirt"). If he would understand that he comes from dirt, as if from the earth, which is the lowest of the elements, he would eagerly strive to keep in mind his final resting place, and humbling himself he would consider himself the lowest. Hence, the Psalmist implored, "Let the peoples know that they are human beings."[27] Hence, when he opened the eyes of the man born blind, the Lord expressly "anointed his eyes with mud," John 9.[28] The man blind from birth appropriately designates someone who glories vainly in his lineage or his gifts of nature and is blinded. If he places dirt before the eyes of his mind, i.e., the vileness of the beginning and end of human beings, he will grow humble by suppressing the pompousness of a blinding pride and he will see. In a letter to King Alexander, Dindimus, king of the Brahmins, wrote as follows: "We are warmed by the sun, we receive moisture from the dew, we quench our thirst from a brook, the ground serves as our bed, anxiety does not interrupt our sleep, thinking does not weary our mind. We have no grand designs to rule over human beings like ourselves, nor do we demand servitude from anything besides our bodies, which we value only as the servants to our souls. And besides, it is an act of barbarity to force people into compliance when the same nature gave birth to them as to us as brothers, just as they, too, are promised an inheritance from the one God, the father of all good things*."[29]

24 Gregory the Great, *Reg. past.*, 1.11, PL 77:25B.

25 Tob 2:10.

26 Cf. Hugh of Saint-Cher (?), *Expositio super Apocalypsim*, 16, in Thomas Aquinas, *Opera omnia* (Parma, 1852–1873), available at http://www.corpusthomisticum.org/x1a15.html (accessed December 10, 2010), and William Peraldus, *Sum. de vit.*, 6.2.7, vol. 2, fol. 126ra. In both texts the statement about pride is also attributed to Gregory. It is John Cassian who notes in *Conlationes xxiiii*, 5.10, ed. Michael Petschenig, CSEL 13 (Vienna, 1886), 130, that monks are in danger of laying themselves open to vainglory and pride precisely when they are victors in the battle against vices.

27 Ps 9:21.

28 John 9:6.

29 Cf. *Epistola Dindimi prima ad Alexandrum*, ed. Pfister, in *Kleine Texte zum Alexanderroman*, 12.

On the second distinction, namely envy.[30] *On the unhappy eye of the envious person pining away at the good fortune of his fellow human being**.

The eye of the envious person, like the eye of someone who squints, does not correctly see the person whom it envies, and for this reason it is said in 1 Samuel 18 that after David had conquered the Philistines, "Saul did not look at David with favorable eyes from that very day onward."[31] Someone else's happiness is torture to the envious person, for:

> The envious person pines away at the prosperity of another,[32]

and for this reason he is called envious (*invidus*), as if he were "not seeing" (*non videns*). Indeed, when jealous people envy someone else's happiness, they cannot look at others' preferment or prosperity. In Genesis 31 it is said: When Laban heard that Jacob "was renowned for his riches," he did not turn his face to him "as [he had] yesterday and the day before yesterday."[33] And in Genesis 28 it is said: Esau saw that "his father did not willingly look at the daughters of Canaan,"[34] and so on, Canaan being interpreted to mean "possessing."[35] Namely, envy is not directed toward the dispossessed, but toward those in possession of many goods. Aristotle says that envy is a sadness at the visible prosperity of a good person.[36] And

30 For the outlines of envy in the Middle Ages, see F.N.M. Diekstra, "The Art of Denunciation: Medieval Moralists on Envy and Detraction," in *In the Garden of Evil*, ed. Newhauser, 431–54; Mireille Vincent-Cassy, "L'envie au moyen âge," *Annales. Économies, Sociétés, Civilisations* 35 (1980): 253–71.

31 1Sam 18:9.

32 Horace, *Ep.*, 1.2.57, ed. Wickham, 2nd ed., ed. Garrod (cf. Walther, *Proverbia*, no. 12788).

33 Gen 31:1–2.

34 Gen 28:8.

35 Jerome, *Lib. interp. Hebr. nom.*, Gen., ed. de Lagarde, CCSL 72:61.

36 Aristotle, *Topica*, 2.2, trans. Boethius (rec. alpha), ed. Lorenzo Minio-Paluello, Aristoteles Latinus, 5.1–3, Corpus Philosophorum Medii Aevi, 1.5.1–3 (Bruges, 1969), 33. Cf. John of Salisbury, *Policr.*, 7.24, ed. Webb, vol. 2, 211, where Webb's emendation is necessary in a text that misrepresents the phrase *boni alicuius micium* (of a good and gentle person), or its equivalent, that apparently underlies the passage. This is, in fact, the phrase found in the Latin translation of Aristotle and in Peter's work in *P7P8*, whereas *Pr1* has simply dropped the final, and often misunderstood, word. The text of the passage in the *Policraticus* printed in PL 199:702A ("[invidia est] tristitiae ex apparenti prosperitate alicujus initium" – [envy is] the beginning of sadness at someone's visible prosperity) demonstrates how difficult the excerpt from Aristotle proved to be to others, as well, and how often *mitium* was misconstrued as *initium*.

therefore there is no way of life that is effective for you to avoid envy unless you become very wretched, for wretchedness alone is without envy. For this reason Socrates said to his disciple, Plato, who was asking him how he could avoid the envy of his fellow pupils, "Be like Thersites."[37] That is the reason people who are of more distinguished worth are more subject to envy and slander. For the envious, as Boethius says, are unable to lift up their eyes to the light of manifest truth, because they are accustomed to darkness, like birds whose ability to see is brightened at night and dulled by day.[38] And therefore it happens that servants revile their lords with slanderous words, and congregations their prelates. And it rarely happens that a prelate or a ruler is not disparaged by those beneath him, however much the prelate may have been carrying out his office correctly in the meantime.

Thus, the eye of the envious is tortured by the good fortune that it sees in its neighbors, Job 24: "If suddenly the morning dawns," i.e., the beginning of someone else's good fortune, "they believe it is the shadow of death,"[39] i.e., they mourn, as if they could see death hanging over them. And thus Socrates says, "If only the envious would have eyes and ears in every city, so they could be tortured by the happiness of all people!"[40] And someone wanting to invoke the height of all evils on another person said:

> May you envy everyone, envious man, but no one envy you.[41]

This is the person about whom Ecclesiasticus 14 says: "An evil eye for evil things."[42] For it always looks at the misfortune of a neighbor and turns

37 Thersites is mentioned by Homer in book 2 of the *Iliad* as a figure of ridicule who is hit by Ulysses when he makes disparaging remarks about Agamemnon and Achilles; in book 10 of the *Republic*, Plato notes that the soul of Thersites clothes itself in the body of an ape. The point of the *exemplum* is that only the most miserable are not envied. The passage from "Aristotle says" through "Thersites" is adopted from John of Salisbury, *Policr.*, 7.24, ed. Webb, vol. 2, 211–14.

38 Cf. Boethius, *Phil. cons.*, 4.pr.4.27, ed. Bieler, CCSL 94:75.

39 Job 24:17.

40 The saying is attributed to Socrates in William Peraldus, *Sum. de vit.*, 7.1.11, vol. 2, fol. 193ra, and Ps.-Caecilius Balbus, *De nugis philosophorum*, Codex Monacensis, 10.2, ed. Eduard Woelfflin, in *De nugis philosophorum quae supersunt* (Basel, 1855), 22. See also Vincent of Beauvais, *De morali principis institutione*, 21, ed. Robert J. Schneider, CCCM 137 (Turnhout, 1995), 108.

41 Martial, *Epigrammata*, 1.40.2, ed. Wallace M. Lindsay, 2nd ed. (Oxford, 1929) (cf. Walther, *Proverbia*, no. 20162).

42 Sir 14:10.

away from his good fortune, and therefore the envious are correctly designated by Eli, about whom it is said in 1 Samuel 3 that "his eyes grew dim because they could not see the lamp of God before it went out."[43] Such are the envious, blinded by envy: They are unable to see good people, who are "lamps in the house of the Lord,"[44] until they have been extinguished by death, for:

> With death, jealousy subsides.[45]

For some people praise a man after his death whom they envied while he was alive. In this way, the poet also says:

> For that man's splendor is galling whose reputation burdens the ears
> Of those beneath him; deceased, he will be admired.[46]

When a lamp is extinguished, it gives off a bad odor. Thus, the envious person cannot see the lamp until it has been extinguished and stinks, since the envious person, who could not see his neighbor while he was in the light of grace, sees him for the first time when he has been extinguished and has cooled off through sin. For this reason Richard of Saint-Victor says: "The envious man envies the good that he sees; the good that he hears, he hates; the good that he touches, he extinguishes."[47] The envious man strives so hard to achieve another person's evil that he would suffer a serious loss so that the other person would be afflicted by an even more serious one. He is like the person who chose to tear out one of his eyes so that his neighbor would be deprived of both of his.[48]

43 1Sam 3:2–3.

44 For this phrase (*lucerna in domo Domini*), cf., among other possible sources that Peter would have known, the beginning of Pope Alexander IV's bull on the university in Paris, "Universis magistris et scolaribus Parisiensibus" (14 April 1255), ed. Charles Bourel de La Roncière, Jean de Loye, and Auguste Coulon, in *Les registres d'Alexandre IV; recueil des bulles de ce pape publiées ou analysées d'après les manuscrits originaux des archives du Vatican* (Paris, 1902), vol. 1, 96.

45 Ovid, *Amores*, 1.15.39, ed. Kenney, 36 (cf. Walther, *Proverbia*, no. 20803).

46 Horace, *Ep.*, 2.1.13–14, ed. Wickham, 2nd ed., ed. Garrod (cf. Walther, *Proverbia*, no. 32267a).

47 Helinand of Froidmont, *Sermones*, 9, PL 212:556B.

48 This well-known *exemplum* is told concerning an avaricious and an envious man who are granted the fulfillment of a wish. The avaricious man wants the double of whatever the envious one desires; knowing that, the envious one wishes for the loss of one eye. See Avianus, *Fabulae*, 22, ed. and trans. J. Wight Duff and Arnold M. Duff, in *Minor Latin Poets*, LCL 284 (Cambridge, MA, and London, 1934; reprint 1968), 714–17. Cf. Tubach, *Index exemplorum*, no. 4087.

Plainly, that person is unfortunate who is made happy only by the misery of his neighbors and for whom his neighbors' happiness is only a source of misfortune. And since the envious man grieves because of the good fortune he sees his neighbor achieve, and he always remains in this state of grief, he can say that verse in Job 17: "My eye lingers in bitterness."[49] And Chrysostom says on John: "The envious person's eye is tortured by grief*."[50] The envious man lives with a continuous death, insofar as he has been separated from him who is life, from whom, as Plato says, all envy has been removed.[51] The nature of the eagle in inspecting its young is said to be this: It places them in the rays of the sun and preserves and feeds like its own offspring each one that looks at the sun without being harmed, but the eagle throws out as if it were not its own kind those it sees averting their eyes from the sun's rays.[52] So also God himself, who is compared to the eagle in Deuteronomy 32: "Like an eagle calling forth its young to flight and hovering over them,"[53] and so on. Inasmuch as they are unlike him, God throws out the envious* who cannot bear, and are unable to look at, the gleam of good fortune in others. And since they refuse to see the light of truth, they will fittingly receive the darkness and misery of hell.

On the third distinction, namely wrath.[54] On the agitated eye of the angry man, who is temporarily insane.*

Wrath agitates the eye of reason, and because of this the Psalmist says, "My eye has been agitated by madness."[55] And elsewhere: "My eye has been agitated by wrath."[56] He calls wrath madness, for according to

49 Job 17:2.

50 Cf. John Chrysostom, *Commentarius in sanctum Joannem Apostolum et Evangelistam*, Hom. 55.3, PG 59:305. The quotation from Chrysostom is also found in the same wording as in Peter's text in the anonymous *Liber Pharetrae*, 2.8, printed in Bonaventure of Balneoreggio, *Opera omnia*, ed. Adolphe C. Peltier (Paris, 1866), vol. 7, 72.

51 Cf. John of Salisbury, *Policr.*, 2.12, in *Policraticus I–IV*, ed. Keats-Rohan, CCCM 118:91.

52 Cf. Hugh of Fouilloy (Ps.-Hugh of Saint-Victor), *Aviarium*, 2.60, ed. and trans. Clark, 250–51.

53 Deut 32:11.

54 On aspects of medieval wrath, see *Anger's Past: The Social Uses of an Emotion in the Middle Ages*, ed. Barbara Rosenwein (Ithaca and London, 1998).

55 Ps 6:8.

56 Ps 30:10.

Seneca someone who has been angered is insane for a time,[57] and the poet says:

> Wrath is a short-lived madness.[58]

For this reason Hugh of Saint-Victor says: "Pride robs a person of God, envy of his neighbor, wrath of the person himself."[59] Moreover, an angry man's internal agitation is occasionally made visible in his external eye, and so describing the state of the wrathful person, Architrenius* says:

> And since the inner peace of his soul is disturbed, he thunders out
> His words, his eye glints, his facial expression is threatening,
> He puffs up his physical frame, hurling and disgorging wrath from his eyes,
> And he spits out vomit from a fuming mouth.[60]

It has also been proven that some people's eyes are by their nature infectious* if they look at someone when they happen to be in a state of violent anger. For this reason, Solinus relates in the *Wonders of the World* that certain women are found in Scythia who are called *bicie* or otherwise *sicie*.[61] And he asserts that they have double pupils in their eyes and that they kill with a glance if they happen to look at someone when they are violently angry.[62] And since the eye of reason is disturbed in the wrathful person, he cannot correctly judge matters, nor is this surprising since he has been spattered with the blood of wrathfulness. For wrath according to Aristotle is a boiling of the blood surrounding

57 Cf. Seneca, *De ira*, 1.1.2, ed. Leighton D. Reynolds, in *Dialogorum libri duodecim* (Oxford, 1977), 39.

58 Horace, *Ep.*, 1.2.62, ed. Wickham, 2nd ed., ed. Garrod (cf. Walther, *Proverbia*, nos. 12869, 12870, and 12870a).

59 Hugh of Saint-Victor, *De sacramentis Christianae fidei*, 2.13.1, PL 176:525C; Richard of Saint-Victor, *Expositio moralis in Abdiam*, in fine, PL 175:401C.

60 Johannes de Hauvilla, *Arch.*, 5.6.130–31, 5.5.105 (first half of the line), 5.6.142 (second half of the line)–43, ed. Schmidt, 200; trans. Wetherbee, 123, 125. Note, however, that in the *Architrenius* these chapters describe the personification of pride (chapter 5) and the proud monk (chapter 6).

61 In *Pr1* the phrase *"bicie" alias "sicie"* is the result of scribal dittography. *P7P8* has simply *bicie.*

62 Cf. Gaius Julius Solinus, *Collectanea rerum memorabilium*, 1.101, ed. Theodor Mommsen, 2nd ed. (Berlin, 1958), 26.

the heart.[63] And hence it is said in Job 17: "My eye has become dim-sighted with indignation."[64] And just as an image does not appear in water which is stirred up, so also truth does not shine out in the mind when it has been stirred up through wrath, since according to the poet:

> Wrath hinders the heart so it cannot discern what is true.[65]

The devil takes care to achieve this disturbance, for he is like a skilled fisherman who stirs up the water in order to catch a fish. For the devil knows that if he disturbs someone's eye of reason, he will capture that person and lead him off willingly, on the example of the horse which walks around the millstone only when its eyes have been covered up.

Hence it also happens that the angry person, whose internal eye has been stirred up by the violent impulse of wrath, occasionally imposes an excessive punishment. For this reason Valerius Maximus relates in book 4 of the *Memorable Sayings and Deeds* that when Archytas of Tarentum had studied philosophy for a long time under Pythagoras in distant parts, he at last returned home to his native country after having obtained the gift of knowledge. When he saw that his estates had been ruined and wasted through the negligence of an overseer, he is reported to have spoken this notable sentiment while contemplating the man who had behaved so badly: "I'd inflict punishment on you right now if I weren't furious with you. For I'd rather dismiss you unpunished than punish you far more harshly than is just because of my wrath."[66] Jerome tells the same story to the widow Salvina*, where he relates that Archytas said, "I would've summoned you already for lashes if I hadn't been enraged."[67] Valerius also relates in the book just referred to that when Plato had become violently enraged over an offense by his servant and was afraid that he could not observe moderation in his punishment, he entrusted the decision about the man's correction to his friend Speusippus, for Plato considered that it would be disgraceful for him to bring it about

63 Aristotle, *De anima*, 1.1 (403a29), ed. David Ross (Oxford, 1961); trans. John A. Smith, in *The Works of Aristotle Translated into English*, ed. William D. Ross (Oxford, 1931; reprint 1963), vol. 3. Aristotle is describing how the natural scientist, as opposed to the dialectician, would define anger.

64 Job 17:7.

65 *Disticha Catonis*, 2.4.2, ed. Marcus Boas, 2nd ed., ed. Hendrik Johan Botschuyver (Amsterdam, 1952), 101 (cf. Walther, *Proverbia*, no. 11566).

66 Valerius Maximus, *Fact. et dict. mem.*, 4.1.Ext.1, ed. Briscoe, vol. 1, 233–34 (cf. Tubach; *Index exemplorum*, no. 3991).

67 Jerome, *Ep.*, 79.9.4, ed. Hilberg, CSEL 55:98 (cf. Tubach, *Index exemplorum*, no. 3991).

that a misdemeanor by a servant and his punishment of it would deserve* equal censure. For this reason Valerius says: "Thus, Archytas' moderation was too generous; that of Plato more temperate."[68] Therefore, it is said in Job 36*: "Do not let wrath best you so that you overwhelm someone."[69] And Cicero says in book one of *On Duties*: "You should beware of wrath while punishing someone, lest the punishment be greater than the offense."[70] From what has been said we can elicit a very beneficial lesson, that when someone is in a state of wrath, he ought not to undertake anything, but rather should put it off for more level-headed consideration in the future when the disturbance of wrath has been quieted down, according to the verse of poetry:

> While fury is flowing, withdraw from the furious current.[71]

For this reason one reads in the *History of the Romans* that because the emperor Theodosius was very likely to issue harsh commands, he was chastised in the following way by a most famous philosopher: When the philosopher observed that the emperor "was too easily roused, to prevent him from imposing a harsh sentence the philosopher advised him to run through the twenty-three letters of the alphabet in his memory when he began to grow angry so that, with his mind occupied by something else, the arousal of his passion would die down."[72] By acting in this way, if he imposed anything at all, it would be just. For this reason as Seneca says:

> The law sees an irate man; he does not see the law.[73]

68 Valerius Maximus, *Fact. et dict. mem.*, 4.1.Ext.2, ed. Briscoe, vol. 1, 234 (cf. Tubach, *Index exemplorum*, no. 3991).

69 Job 36:18.

70 Cicero, *De off.*, 1.25.89, ed. Winterbottom, 36.

71 Ovid, *Remedia amoris*, 119, ed. Edwin J. Kenney, in *Amores, Medicamina faciei femineae, Ars amatoria, Remedia amoris*, 2nd ed. (Oxford, 1994), 230 (cf. Walther, *Proverbia*, no. 6541).

72 Paul the Deacon, *Historia Romana, Libri XI–XVI*, 12.5, ed. Hans Droysen, in *Eutropi Breviarium ab urbe condita cum versionibus Graecis et Pauli Landolfique additamentis*, MGH, Auctores antiquissimi, 2 (Berlin, 1879), 191; reprint as *Pauli Historia Romana*, MGH, Scriptores rerum Germanicarum in usum scholarum separatim editi, 49 (Munich, 1978), 98. In Paul's text, it is the 24 letters of the *Greek* alphabet that are to preoccupy the mind as a way to prevent wrathful action.

73 Publilius Syrus, *Sententiae*, L.304, ed. Otto Friedrich (Berlin, 1880; reprint Hildesheim, 1964), 54; *Sententia communis* (Walther, *Proverbia*, no. 13717).

And Aristotle says that two things are extremely adverse to deliberation, namely haste and wrath.[74] Hence, we can also learn that no one ought to be indignant if a wrathful person strikes out at him, since the angry person is blind, but rather someone who is not on his guard against the wrathful person ought to be angry* at himself. Therefore, Seneca says: "'People are saying bad things about you.' I'd be upset," he says, "if they were acting rationally, but they suffer from a disease."[75]

On the fourth distinction, namely spiritual sloth.[76] *On the putrid eye of the slothful person, lazy and drowsy when it comes to his own salvation*.*

The slothful, whose internal eyes have been weighed down with the sleep of laziness, are signified by the sleeping disciples, about whom it is said in Matthew 26: "He found them sleeping,"[77] for their eyes were heavy. They can say this verse by the Psalmist: "My spirit slept with weariness."[78] This event took place because of a lack of reflection on eternal abundance. If they had paid attention to the promised wealth of heavenly glory, they would have banned sleep from their minds and kept their eyes open and focused on obtaining this wealth, Proverbs 20: "Keep your eyes open, and you will eat your fill of bread."[79]

The sleep of spiritual sloth is also caused by a lack of reflection on future punishment. What Jacob said in Genesis 31 can be used to explain this sleep: "I was driven on by heat during the day, and by frost at night,

74 William Peraldus, *Sum. de vit.*, 8.1.2, vol. 2, fol. 196rb (attributed here to Aristotle); cf. Ps.-Caecilius Balbus, *De nug. phil.*, Codex Monacensis, 28.5 (no attribution); and Codices Parisini, 5 (attributed to Phocion), and 55b (attributed to Socrates), ed. Woelfflin, 29, 38, 41. The saying is also attributed to Socrates in Iacopo da Varazze (Jacobus de Voragine), *Chronica civitatis Ianuensis*, 8.1, ed. Giovanni Monleone in *Iacopo da Varagine e la sua Cronaca di Genova dalle origini al MCCXCVII*, Fonti per la storia d'Italia pubblicate dal R. Istituto Storico Italiano per il Medio Evo. Scrittori, secolo XIII, [85] (Rome, 1941), vol. 2, 155.

75 Ps.-Seneca, *Ad Gallionem de remediis fortuitorum liber*, 7.2, ed. Friedrich Haase, in *Opera quae supersunt* (Leipzig, 1886), vol. 3, 450.

76 *Accidia* (and its derivatives) is the technical term, translated here as "spiritual sloth," used earliest in monastic teachings on the capital vices. For its background, see Michael Theunissen, *Vorentwürfe von Moderne: Antike Melancholie und die Acedia des Mittelalters* (Berlin and New York, 1996); Siegfried Wenzel, *The Sin of Sloth: Acedia in Medieval Thought and Literature* (Chapel Hill, 1967).

77 Matt 26:40.

78 Ps 118:28.

79 Prov 20:13.

and sleep fled from my eyes."[80] That person is driven on by heat and frost who observes with the eyes of his mind the punishments of hell, where the damned will cross "from the snowy waters to the immoderate heat," Job 24,[81] and if he reflects on these punishments sleep flees from his eyes, since he does not slumber in the sleep of laziness. Macrobius relates in the *Saturnalia* that when Emperor Augustus was told about the huge amount of debt one of the Roman Equites had concealed while he was alive, he immediately ordered someone to bring for his own use the mattress on which that knight had habitually slept. And to his attendants, who were wondering about it, he gave this reason for his command: "For my own sleep I must acquire and carefully inspect the mattress on which that man was able to sleep who had so many debts."[82] Thus, one must marvel even more how the eyes of sinners, understanding they are liable for a debt of eternal punishment, still long for sleep, since a reflection on punishment ought to open their eyes that sin was able to close. For this reason Job said after his scourging, "Now my eye sees you," Job 42.[83] And in Jeremiah 1 it is said: "I see a rod keeping watch."[84] The rod keeping watch against the sleep of the body and spiritual sloth is the contemplation of future punishment. And therefore one reads in the *Lives of the Fathers* that Abba Achilles responded as follows to one of the brothers when he was being troubled by sloth and asked him what he should do: "You suffer from spiritual sloth because you haven't yet witnessed the rewards we hope for or the punishments we fear, for if you were to examine these matters diligently, you'd lie down in your cell without the sluggishness of sloth even if it were full of worms."[85]

This sluggishness of spiritual sloth makes not only the internal eye of the soul sleep with laziness, but also the external eye. For there are some churchmen who sleep in their bed chambers during the hour for the morning office, or fall asleep in the choir of the church, if they even happen to be present. Against them, the blessed Bernard in his *Letter to the Cluniacs*: "What place does idleness have here, oh good soldiers? Your

80 Gen 31:40.

81 Job 24:19.

82 Macrobius, *Saturnalia*, 2.4.17, ed. James Willis (Leipzig, 1963), 145–46. The Equites, roughly speaking, were Roman equivalents to the later knightly class of the Middle Ages.

83 Job 42:5.

84 Jer 1:11.

85 *Vit. pat.*, 3.107, PL 73:780D–81A; cf. ibid., 5.7.28, PL 73:900C. Abba Achilles was one of the Desert Fathers, anchoritic monks who lived in the desert in North Africa beginning in the third century; several sayings are attributed to him in the alphabetical collection of the sayings of the Desert Fathers.

comrades are wheeling about in blood and gore, but you are either enjoying a voluptuous meal or taking your morning naps. Others 'are using their time to the full, since the days are evil,'[86] but you on the other hand use up the long nights in sleeping and celebrate your lazy days with telling empty stories."[87] The same author on the Song of Songs: "It saddens me that some of you are sunk in a heavy sleep during the night office, that you do not revere the citizens of heaven but in the presence of the heavenly princes you appear like corpses, even though they are very moved when we show ardor and delight at being present during our religious services. I am afraid they might depart in indignation at some point, hating our laziness."[88]

And if sleep of this kind or slumber at such a time can be reasonably censured in any clergyman, then all the more so in a prelate who ought to keep watch carefully like a sentinel of the church so that "he neither slumbers nor sleeps who watches over Israel."[89] For he should guard "his flock through the watches of the night"[90] like an ever watchful shepherd. Helinand, monk of Froidmont, relates among other matters a story of this very kind* about himself in his *Chronicles*, book 8: "Listen," he says, "to a fable; not a fable, but a real event! Philip, bishop of Beauvais, stayed with us once as a guest, not like one motivated to devour our goods, but out of devotion (for the well-known hospitality of our order had already attracted people who wanted to devour it, more than its remarkable holiness had found people who honored it). Now, the bishop I just mentioned instructed me to see to it that he would hear morning mass. When I came to him the next day after Prime had already been sung, I found him still asleep, and no one in his household dared wake him. But I went to him on my own and woke him up, saying as if it were a joke, 'A long time has already passed since the sparrows arose to praise the Lord, and our bishops are still snoring in bed. Listen, father, to what the Psalmist says: "My eyes have come to you before dawn."[91] Concerning this verse Ambrose's *Gloss* says: "It is unbecoming for a Christian if a ray of sunshine finds him lazy in bed."'[92] The bishop was awakened by this speech, thrown into a state of confusion, and was angry with me because

86 Eph 5:16.

87 Bernard of Clairvaux, *Apologia ad Guillelmum abbatem*, 9.21, ed. Jean Leclercq and Henri M. Rochais, in *Opera* (Rome, 1963), vol. 3, 99–100.

88 Bernard of Clairvaux, *Serm. sup. Cant.*, 7.4.4, ed. Leclercq, Talbot, and Rochais, in *Opera*, vol. 1, 33.

89 Ps 120:4.

90 Luke 2:8.

91 Ps 118:148.

92 Cf. *Gloss. ord.*, Ps. 118:148, ed. Rusch, vol. 2, 620a in margin.

I had censured him so boldly. Being rather annoyed he said, 'Go away, wretch, and pick off your lice.' But I turned his agitation into a joke and responded immediately, 'See to it yourself, father, that your worms don't kill you, for I've already killed my worms. You should know, father, that there is this difference between the worms of the rich and those of the poor: The worms of the rich kill the rich, but the worms of the poor are killed by the poor. Read the stories of the Maccabees, of Josephus, and in the Acts of the Apostles, and you'll find that the most powerful kings, Antiochus and Herod Agrippa, were consumed by worms.'[93] By this reasoning and appeal to an authoritative source, the bishop was overwhelmed and fell silent at once."[94]

Thus, the slothful have internal eyes that have been darkened, and so it is said in Proverbs 10: "As smoke to the eyes, so is the indolent person to those who sent him."[95] For as smoke darkens the eye of the body, so does spiritual sloth the eye of the mind. Nor is it any wonder that the eyes of the slothful, who are sluggish with laziness, become darkened so that they do not see, for sight is supported by taste. For this reason it is said in 1 Samuel 14: "Jonathan stretched out the rod which he was holding in his hand and dipped the tip of it in a honeycomb and raised it to his mouth and his eyes became bright."[96] The straight branch signifies a simple and upright intellect; it is held in the hand when it is made use of in deeds. The tip of the rod is the subtlety of intelligence. This is dipped into a hon-

93 Cf. 2Mac 9:9 (Antiochus IV Epiphanes) and cf. Acts 12:23 (Herod Agrippa). Flavius Josephus, *The Jewish War*, 1.39, ed. and trans. Henry St.J. Thackeray, in *Josephus*, 2, LCL 203 (Cambridge, MA, and London, 1927; reprint 1956), 22–23, reports only that Antiochus IV Epiphanes died after having been defeated by Judas Maccabaeus. Of the death of Herod Agrippa, Josephus says in *Jewish Antiquities*, 19.346–50, ed. and trans. Louis H. Feldman, in *Josephus*, 9, LCL 433 (Cambridge, MA, and London, 1965), 378–81, only that Agrippa died with acute abdominal pain. Agrippa's grandfather, however, Herod the Great, is said by Josephus to have died from an illness accompanied by intestinal pain and an inflammation that produced worms: *Jewish Antiquities*, 1.656, ed. and trans. Henry St.J. Thackeray, in *Josephus*, 4, LCL 242 (Cambridge, MA, and London, 1930), 310–13.

94 The entire passage from "Helinand, monk of Froidmont" to "fell silent at once" is taken from Helinand of Froidmont, most likely via Vincent of Beauvais, a text that is verbally very close to Peter's work, though Vincent's text leaves out the references to the Psalmist and Ambrose's *Gloss*. These may be interpolations, but are nevertheless part of the authoritative transmission of Peter's text. Cf. Vincent of Beauvais, *Spec. hist.*, 29.116, in *Speculum quadruplex*, vol. 4, 1225a–b; *Flores a Vincentio Bellovacensi collecti*, 1.9, PL 212:730C–31A. Philip of Dreux (1158–1217) was both a crusader and the bishop of Beauvais.

95 Prov 10:26.

96 1Sam 14:27.

eycomb when it arrives at the hidden sweetness of Holy Scriptures. The hand is held to the mouth when a deed is combined with religious teaching. And the spiritual food is the accomplishment of a meritorious act, as the Truth says in John 4, "My food is to carry out the will" of my father "in order to complete his work."[97] Thus, since the lazy do not eat of this food, their eyes are darkened, and they can say that verse of the Psalm: "My eyes have grown dim with the lack of food."[98] Therefore, it is said in Zechariah 11: "His arm will wither with dryness and his right eye will grow dark and be blinded."[99] In the arm one understands a meritorious act. Thus, the arm withers with dryness when a person is sluggish with laziness and does not carry out a good deed, and then by necessity the right eye grows dark since the intellect is blinded in terms of a knowledge of spiritual matters. And therefore Solomon, whose eyes above all others were illuminated by wisdom, began to grow lethargic from laziness after the temple of the Lord had been constructed; from that moment on he became infatuated with idolatry, as one reads in 1 Kings 11.[100] Thus, the slothful person is an unfortunate one, losing his interior eye while he pursues a life of idleness.

It is related in the manner of a parable that a certain king had three sons. Desiring to make arrangements for his kingdom, he had his three sons called together before him and said that he wanted to bequeath his kingdom to the laziest one. Then the first one said, "The kingdom, in that case, ought to be mine, since I'm so lazy that when I'm sitting by the fire I allow my leg to be burnt rather than pull it away from the heat of the fire." "On the contrary," said the second one, "I'm better able to rule since I'm lazier, for even if I were to have a rope around my neck and were at the point of being hanged in the next moment and were to hold a sword in my hand, because of my laziness I wouldn't stretch out my hand to cut the cord." "On the contrary," responded the third, "I ought to rule; I surpass the others in laziness. For when I lie on my back in bed and water is dripping on both of my eyes, on the right one and the left one, because of my utmost laziness I don't move off the bedspread to the right side or the left." The king bequeathed his kingdom to the third son, considering him more completely lazy. – This king stands for the devil who is "king over all the sons of pride," Job 41.[101] By the first of the three sons a person is fittingly designated who keeps company with an evil crowd, because of

97 John 4:34.
98 Ps 87:10.
99 Zech 11:17.
100 Cf. 1Kgs 11:4–11.
101 Job 41:25.

whom he is incited to evil. He chooses rather to be burnt up by the fire of sin than to be separated from this kind of companionship. But the second son stands for a person who sees that he is tied up or even bound everywhere by the ropes of his sins and that he is going to be hanged by them in the next moment on the gallows of hell, and nonetheless he is too lazy to cut through the rope of sin with the sword of his tongue, i.e., by confession. Water drips on both of the third son's eyes, that is to say on the right one and the left one, and he fittingly designates someone who hears the teaching on the joys of heaven and the torments of hell, and nevertheless due to laziness he does not change his position. He wants to be neither on the right side, out of love and desire for rewards, nor on the left, so that at least he forsakes sin terrified by the fear of torments, and so on.[102]

The fifth distinction follows, namely avarice.[103] On the eye of the avaricious person blinded by earthly possessions.*

The eyes of the avaricious are strewn with the dust of worldly possessions and therefore they have been blinded, for "the eyes of fools are on the ends of the earth," Proverbs 17.[104] And about the verse of the Psalm: "In the morning I will stand by you and I will see,"[105] the *Gloss* says: "He does not see who hangs onto temporal matters."[106] And hence they are like moles, deprived of eyes but nevertheless bearing the marks of eyes, that dig up the earth and throw it on top of themselves.[107] So also the avari-

102 Cf. *Gesta Romanorum*, 91 (83), ed. Hermann Oesterley (Berlin, 1872), 418–19; Robert Holcot, *Liber moralizationum historiarum*, 47, in *In Librum Sapientiae Regis Salomonis Praelectiones CCXIII ...* (Basel, 1586), 748. Cf. Tubach, *Index exemplorum*, nos. 2896 and 3005. For a comparison of the versions of the *exemplum* in the *Gesta Romanorum*, Peter's work, and Holcot's that points to Peter's treatise as a witness for an early form of the transmission of the *Gesta Romanorum*, see Newhauser, "Der 'Tractatus moralis de oculo' des Petrus von Limoges und seine *exempla*," 113–17.

103 On the background of avarice in the Middle Ages, see Richard Newhauser, *The Early History of Greed: The Sin of Avarice in Early Medieval Thought and Literature*, Cambridge Studies in Medieval Literature, 41 (Cambridge, Engl., 2000; reprint 2006); Alexander Murray, *Reason and Society in the Middle Ages* (Oxford, 1978), chapter 3; Lester K. Little, "Pride Goes before Avarice: Social Change and the Vices in Latin Christendom," *The American Historical Review* 76 (1971): 16–49.

104 Prov 17:24.

105 Ps 5:5.

106 *Gloss. ord.*, Ps. 5:5, ed. Rusch, vol. 2, 461b interlinear.

107 Cf. Tubach, *Index exemplorum*, no. 3315.

cious in this world hold onto something that bears a resemblance to prudence, but they do not possess the truth of wisdom, and as long as they seek worldly things they are burying themselves alive in worldly desires, since as it is said in Proverbs 12: "The faithless person's desire is his tomb."[108] This is the eye which is spoken about in Zechariah 5: "Their eyes are on the entire earth,"[109] since the full intent of the avaricious is focused on the acquisition of earthly goods throughout the entire earth.

The eye cannot see an object plainly and clearly unless the object is separated from it by a proportional distance,[110] and therefore since greedy people put temporal wealth over the eyes of their heart, they form a perverse judgment about riches, so that they consider things that are worthless to be precious. And just as a corporeal eclipse will occur because of the interposing of an object that casts a shadow between the sun and the eye of the body, so from the fact that through a too ardent love something earthly is placed between the eye of the heart and the sun of justice, a spiritual eclipse will take place in the soul of the greedy person. The Psalm: "Fire fell upon them," namely of burning avarice, "and they did not see the sun," add here: of justice.[111] And therefore fittingly silver (*argentum*) is understood as "the ardent desire of nations" (*ardor gentium*). Just as a bear is blinded by a glowing basin-shaped lamp, so is the greedy person by the love of worldly things.[112] And since wherever your love is, there is your eye,[113] the greedy person directs the eye of his intention wholly towards gold, which he loves most. And while he delights in its shininess and beauty, he is blinded spiritually in the manner of a bear. Since avarice dulls the sharpness of a person's vision, it makes him grind at the millstone, i.e., at the revolving mutability of temporal matters, as one reads concerning Samson in Judges 16: "Whom, when the Philistines had seized him, immediately they gouged out his eyes and forced him to grind grain locked up in prison."[114]

Nor is it surprising if the greedy person is unable to see clearly since he has "a beam in his eye."[115] For a person who yearns for an ecclesiasti-

108 Prov 12:12.

109 Zech 5:6.

110 On the need for a proportionate amount of distance between the eye and the visible object for vision to occur, see below, chapt. 11.3, p. 127.

111 Ps 57:9.

112 Cf. William Peraldus, *De eruditione principum*, 5.10, in Thomas Aquinas, *Opera omnia* (Parma, 1852–1873), available at http://www.corpusthomisticum.org/xreo.html (accessed December 10, 2010) (cf. Tubach, *Index exemplorum*, no. 518).

113 Cf. Matt 6:21; cf. Luke 12:34. See also above, chapt. 7.6, p. 57, and 7.11, p. 71.

114 Judg 16:21.

115 Matt 7:3.

cal office with all his desire has a beam in his eye. A spider hangs from a beam and disembowels itself completely by pulling threads out of its entrails and spinning a useless web. So also the ambitious or greedy person hangs mentally from the honor which he yearns to get and hungers for, and he pulls out threads when he is drawn in different directions by various desires; spins a useless web when he works hard for a transitory good which he sometimes does not even attain, and if by chance he obtains it, it does not remain his for long; and disembowels himself moreover when, consumed by many different thoughts, he weakens the power of his soul. Therefore, the greedy who pursue honors can truly say: "Our years* will be thought of as a spider."[116] And just as in Leviticus 21 someone "having a film over his eye"[117] may be barred from serving in the temple, so also an ambitious person ought to be justly rejected from ecclesiastical benefices. He has, as it were, a film of longing for honor in the church that blinds the eye of his mind. The verse in Jeremiah 22 can be said to him: "Your eyes and your heart are on avarice."[118] And the power of greed to produce blindness is clear from the fact that a prophet blinded the eye of greed, and an apostle corrupted by money betrayed his innocent Lord to be crucified at the hands of the faithless.[119]

Avarice has established a ludicrous and lamentable fellowship with old people, so that when they should need things less, they long for them more passionately, and the quicker they come to the point of relinquishing possessions, the more eager they are to face dangers [to keep them]. For the hand which held out alms in youth is commonly withheld in old age, and the one which was withheld in youth is never or only rarely held out by the aged. However much the avaricious person is deprived of light for his spiritual eyes, after he has gazed on the coins in his safe he nevertheless appears to himself very prudent and having more foresight than all others. For he contemplates what Petronius* says:

> Whoever has money sleeps in a safe shade
> And may direct his fortune at will.
> Let him store up songs, deliver speeches, snap his fingers at everyone
> And win his cases and be better than Cato.
> To be brief: Whatever you want, wish for it with money lying about,
> And it will come. Your safe holds Jupiter locked up inside it.[120]

116 Ps 89:9.

117 Lev 21:20.

118 Jer 22:17.

119 Cf. John 12:40, Isa 6:10 (in reference to Isaiah?); cf. Matt 26:14 (Judas).

120 Petronius, *Satyricon*, 137, ed. Konrad Mueller, in *Satyricon reliquiae*, 4th ed. (Leipzig, 1995), 168–69 (cf. Walther, *Proverbia*, no. 25559).

But the verse in Revelation 3 can be said to him: "You say 'I am wealthy, have been made rich, and need nothing!' and you do not know that you are wretched and miserable and poor and blind and naked":[121] wretched in life, miserable in reputation, poor in good deeds, naked in virtues, and blind, since you are deprived of spiritual understanding.

And since the greedy or avaricious person has been blinded, as was said, by the dust of temporal matters, therefore he is fittingly designated by Tobit, blinded by the excrement of swallows. It is said about him in Tobit 2: "From a nest of swallows hot excrement fell on his eyes and he was blinded."[122] The excrement of swallows stands for temporal possessions and these are said to be hot since they are loved more than they should have been, and therefore those who love them are blinded. But how this vision can be restored is spoken about in Tobit 6: "Split open this fish and preserve for yourself its heart and gall bladder. For these are necessary things which can usefully be made into medicine."[123] The fish is Christ; whoever thinks deeply about the passion of Christ splits this fish open. Whoever considers with what a great heart Christ loves us preserves his heart, and whoever understands how great was the bitterness of Christ's passion preserves his gall bladder*. For these are the necessary items that are used to make medicine for the spiritual eyes. In another way, too, the blindness just mentioned is cured by the gall bladder of a fish, i.e., by thinking about the bitterness that exists in riches. For "there is labor in acquiring them, fear in possessing them, and sadness in letting go of them."[124] If someone considers this diligently, he will not love temporal riches, and thus when he is enlightened spiritually he will regain his eyesight that he lost through greed.

On the sixth distinction, namely gluttony.[125] *On the eye of the gluttonous person wounded by being excessively voracious**.

The eyes of the gluttonous are said to be fleshly since their entire intent is to obtain meat, Job 10: "Do you have eyes of flesh?"[126] And Juvenal says:

121 Rev 3:17.

122 Tob 2:11.

123 Tob 6:5.

124 Pope Innocent III, *De miser. cond. hum.*, 1.14, ed. and trans. Lewis, 117.

125 For an introduction to aspects of medieval gluttony, see Mireille Vincent-Cassy, "La *gula* curiale ou les débordements des banquets au début du règne de Charles VI," in *La Sociabilité à table: Commensalité et convivialité à travers les âges*, ed. Martin Aurell, Olivier Dumoulin, and Françoise Thelamon, Publications de l'Université de Rouen, 178 ([Rouen], 1992), 91–102; and eadem, "Between Sin and Pleasure: Drunkenness in France in the Late Middle Ages," trans. Erika Pavelka, in *In the Garden of Evil*, ed. Newhauser, 393–430.

126 Job 10:4.

> There are those for whom life's exertion is centered on their palates alone.[127]

The eyes of the gluttonous are very sharp and they are, as it were, eagle-eyed in contemplating food. In Job 39 it says: "Will the eagle be lifted up to your teaching and make its nest in the lofty places?"[128] From there it looks out for food and its eyes peer far into the distance. For it is the nature of the eagle that once it has flown up to the heights, from there it catches sight of fish in the sea and descends like a peal of thunder to gulp them down.[129] And the eyes of the gluttonous contemplate food not only in water, but also in all other elements, according to the verse of the poet:

> Among belly worshipers these questions are vexing: Which
> Lake is inhabited by superior fish, which airy region is
> Teeming with birds, which land pours out edible game?[130]

This is the eye which is spoken about in Deuteronomy 33: "The eye of Jacob in a land of grain and wine,"[131] for as the poet testifies concerning the gluttonous:

> Their ingenuity is to search for wine, for the grace of dining
> Is diminished without wine, nor is enjoyment complete
> At a table where water is served, and it is an evil deed
> To shipwreck guests with food alone: Bacchus makes
> Something festive of a meal.[132]

And just as the eyes of spiritual people are always on the Lord, so the eyes of carnal or gluttonous people are always on food and on their stomach, which they cultivate and make their god, Philippians 3: "Whose god is their stomach,"[133] which Hugh explains in book 2 of *The Soul's Cloister*, saying: "Through this similitude the stomach is compared to a

127 Juvenal, *Sat.*, 11.11, ed. Clausen, 134.

128 Job 39:27.

129 Cf. Hugh of Fouilloy (Ps.-Hugh of Saint-Victor), *Aviarium*, 2.60, ed. and trans. Clark, 250–51.

130 Johannes de Hauvilla, *Arch.*, 2.5.184–86, ed. Schmidt, 150; trans. Wetherbee, 43. The text is taken from a description of a debate among gluttons. See also below, near the end of chapt. 8.6, p. 98.

131 Deut 33:28.

132 Johannes de Hauvilla, *Arch.*, 2.8.259–63, ed. Schmidt, 153; trans. Wetherbee, 47. In this passage Johannes describes the search for the best wine to complete a feast.

133 Phil 3:19.

god when it is said, 'Whose god is their stomach.'"[134] For customarily temples are built for gods, altars are erected, ministers are ordained to serve them, animals are sacrificed, incense is burned. And in fact, for the god of the stomach, the temple is the kitchen; the altar, the table; the ministers, cooks; the sacrificed animals, cooked meats; and the smoke of incense, the odor of the delicacies. But as someone says:

> Do not let your stomach be your lord, but live decently!
> In tasting pay heed to: what, how much, how, when,
> Not: hastily, too expensively, excessively, too eagerly, with too much care*.[135]

When a gluttonous person sees delicious food, he is caught by his eye as if through a fish-hook, Job 40: "By his eyes he will capture him as if with a fish-hook."[136] Food is put deceitfully on a hook used to catch fish, but the metal tip lies hidden inside the food. In the same way, delicious food has also been sprinkled externally with a delectable taste that attracts the gluttonous person, but internally a barb lies hidden that wounds the mind spiritually. For this reason Boethius says in book 3, meter 7, of *The Consolation of Philosophy*:

> All pleasure has this quality:
> It goads those enjoying it
> And like the swarming bee
> When it has poured out its sweet honey,
> It flies away and wounds their hearts
> With a very persistent sting.[137]

And since the allurement of gluttony enters into the soul through the eye, therefore the wise man clearly says in Proverbs 23: "Do not look

134 Hugh of Saint-Victor, *De claustro anime libri iv*, 2.19, PL 176:1072D (quoting Phil 3:19).

135 *Sententiae communes* (Walther, *Proverbia*, nos. 18609 [first verse] and 28136 [second and third verse]). The last verse is found in a number of delineations of the five sub-categories of gluttony, where the first category appears as *praepropere* (too hastily). Cf. for example Thomas Aquinas, *Scriptum super Sententiis*, lib. 4, dist. 15, quaest. 3, art. 4, quaestiunc. 3, arg. 2, in *Opera omnia* (Parma, 1852–1873), available at http://www.corpusthomisticum.org/snp4000.html (accessed December 10, 2010). The categories are derived from the examination of gluttony in Gregory the Great, *Mor. in Iob*, 30.18.60, ed. Adriaen, CCSL 143B:1531.

136 Job 40:19.

137 Boethius, *Phil. cons.*, 3.metr.7.1–6, ed. Bieler, CCSL 94:47 (cf. Walther, *Proverbia*, no. 10531a).

at wine when it turns golden-yellow, and when its color glistens in the glass. It goes down gently, but in the end it will bite like a snake, and like a small serpent (*regulus*) it may pour out venom."[138] The basilisk (*regulus*) kills by being seen only once.[139] Now, when the wise man says, "Do not look at wine," he is censuring the voraciousness of modern people for whose gluttony taste alone does not suffice, but they want all of their senses to become inebriated by one chalice. For sight delights in the clearness of the wine, touch in its coolness, taste in its flavor, the nose in its odor, and since there is nothing in wine that could please the sense of hearing, they add song, the lyre, and the timbrel, Isaiah 5: "The lyre and lute and timbrel and flute and wine at your banquets."[140] Thus, gluttony, which in this way takes prisoners by means of the eyes, often destroys the corporeal eyes, since:

> Drinkers who quaff Falernian wine in the middle of the night[141]

prefer to lose their eyes rather than not to be drunk, saying to those who reprove them that wine is more useful than a bad eye. Therefore it is said in Proverbs 23: "For whom is there woe? For whose father is there woe? For whom are there arguments? For whom a pitfall? For whom wounds for no reason? For whom an undermining* of the eyes? Is it not for those who linger over their wine and are eager to drain off goblets?"[142] Just as gluttony sometimes steals away and blinds the eyes of the body, it also does the same thing to the eyes of the heart, Hosea 4: "Fornication, wine, and drunkenness steal away the heart."[143] And the poet says:

> Shameful tomb of reason:
> Drunkenness.[144]

138 Prov 23:31–32.

139 Cf. Isidore of Seville, *Etymologiarum siue originum libri xx*, 12.4.6, ed. Wallace M. Lindsay (Oxford, 1911), vol. 2, who translates the Greek *basiliscus* with the Latin *regulus*, which is the word used in the passage from Proverbs quoted in the preceding sentence. Peter repeats the term apparently in its Isidorean interpretation. Normally, the basilisk is said to kill by sight alone (*solo visu*), not by being seen once (*solo viso*), as the text reads in *Pr1*. On the basilisk, see also below in chapt. 8.7, p. 99.

140 Isa 5:12.

141 Horace, *Ep.*, 1.18.91, ed. Wickham, 2nd ed., ed. Garrod (cf. Walther, *Proverbia*, no. 22106b).

142 Prov 23:29–30. The standard Vulgate text for v. 29 reads *cui suffusio oculorum* (for whom a cataract of the eyes).

143 Hos 4:11.

144 Walter of Châtillon, *Alex.*, 1.171–72, ed. Colker, 16 (cf. Walther, *Proverbia*, no. 26365b).

When a guide dog for the blind sees a bone in the dirt, it turns aside to the bone with great eagerness and drags the blind person along after itself through the filth, and in just the same way after gluttony has blinded a person, it leads him through the filth of sins. And so it happened to the first human beings: "For the woman saw that the tree was good and pleasant to eat and beautiful to the eyes; she picked the fruit of the tree and ate it."[145]

Thus, when they swallow up food and drink, the gluttonous suppress the spirit and are blinded in their rational eyes, and deprived in this way of rational eyes, they become like beasts. In the likeness of cattle, whose entire care revolves around seeking the lushest parts of the pasture, they are eager to have fine foods and to drain off goblets, and consequently using many types of skills, they seek out stimulants for the palate. Let me put it poetically:

> Equally [debated] are the amount of labor and zeal with which dishes are prepared:
> How they are divided up, with what eagerness they are sliced, can
> Enlarge the passion of gluttony, for presentation creates the charm
> And richness of what is prepared, and substance is much supported
> By appearance, for if excessive labor assist,
> The creation will surpass its ingredients, presentation will both add riches
> To the feasts and fan the flames in burning throats.[146]

Now who, as Aristotle says, "possessing any sense of human decency would rejoice in these two pleasures, copulating and eating, which a man has in common with a pig and a jackass?"[147]

The seventh distinction, on lust.[148] *That a woman's eyes are the darts of unchasteness by which many are wounded*.*

The eye of the body is the gatekeeper of the heart which, if it is chaste, has excluded whatever it knows is indecent. If it is unchaste and seeks

145 Gen 3:6.

146 Johannes de Hauvilla, *Arch.*, 2.5.225–31, ed. Schmidt, 151; trans. Wetherbee, 45. The passage is taken from a description of a debate among gluttons. See also above, at the beginning of chapt. 8.6, p. 95.

147 Cf. John of Salisbury, *Policr.*, 8.8, ed. Webb, vol. 2, 276.

148 Of the many recent studies on aspects of lust in the Middle Ages, see especially Pierre J. Payer, *Sex and the New Medieval Literature of Confession, 1150–1300*, Studies and Texts, 163 (Toronto, 2009); John Kitchen, "Cassian, Nocturnal Emis-

whatever can provide pleasure, it rushes around after all of these things and like a bad gatekeeper it allows those to enter who want to and it invites and encourages even those unwilling to enter. One reads in Genesis 39: "His master's wife cast her eyes on Joseph."[149] These are the darts of unchasteness by which many are wounded, no matter how far away they are. Joseph opposed these darts with the chastity of his eyes. For it is written: "Pay no attention to the deceit of a woman," Proverbs 5.[150] And therefore Joseph was not captured by his eyes. But on the other hand, that happened to Holophernes, Judith 10: "When Judith entered into his presence, immediately Holophernes was captured by his eyes."[151]

The basilisk kills birds flying past simply by looking at them;[152] so, too, the appearance of a woman sometimes drags heavenly men down to the depths of lust and thus kills them spiritually. For this reason in Proverbs 7 it is said of a woman decked out in the apparel of a whore: "She has felled many after wounding them and those who are the strongest have been killed by her."[153] Although the basilisk is a short serpent, it kills many by its sight; thus, although something might seem to be moderate in appearance, nevertheless it wounds and kills many. Namely, it seems probable that when a woman looks at a man lustfully, the lustful vapor emanating from her heart rises to her eyes, infecting the woman's visual rays. Thus infected, the rays travel to the man's eyes (assuming that we see by extramission) and when they arrive at the man's eyes and infect them, the infection passes to the man's heart, just as it proceeded from the woman's heart first. For this is the nature of a poisonous ray, as can be seen through a comparison. A rabid dog has poison in its brain, and when it has wounded someone by biting any part of his body, this person suffers afterwards and becomes infected in his brain, just as the dog was also infected in the same part of the body. In this way, the infection of the heart travels to the eyes. And for this reason the poet says:

---

sions, and the Sexuality of Jesus," in *The Seven Deadly Sins*, ed. Newhauser, 73–94; Ruth Mazo Karras, "The Lechery That Dare Not Speak Its Name: Sodomy and the Vices in Medieval England," in *In the Garden of Evil*, ed. Newhauser, 193–205; Pierre J. Payer, *The Bridling of Desire: Views of Sex in the Later Middle Ages* (Toronto, 1993); and Ruth Mazo Karras, "The Latin Vocabulary of Illicit Sex in English Ecclesiastical Court Records," *Journal of Medieval Latin* 2 (1992): 1–17.

149 Gen 39:7.

150 Prov 5:2.

151 Jth 10:17.

152 Cf. *Gloss. ord.*, Isa. 59:5, ed. Rusch, vol. 3, 86a in margin (PL 113:1301D). See also above, chapt. 8.6, p. 97.

153 Prov 7:26.

Like a watchman's lamp, her eye glows with fires.[154]

And according to Aristotle, when a menstruating woman looks in a mirror, she stains and infects it.[155] Likewise, when a woman looks at a man out of lust, sometimes she infects his soul in which the image of God shines out as in a mirror. Just as the active sight of a woman often causes injuries, i.e., by looking, so the passive sight of that same woman also at times commits no less an offense, i.e., when she is looked at. For this reason Quintilian says in the *Book of Cases*: "The path for our vices goes through the eyes into the heart";[156] and in Jeremiah 9: "Death has climbed in through our windows, it has entered our homes."[157] For as it sometimes happens, a thief who is smaller in stature is sent in through a window to open the gate for the larger thieves who enter, plunder the home, and murder the master of the house. In just the same way an illicit look enters in through the window of the eyes to open the door of consent, and then the enemy comes in, plundering the home of its virtues and killing the spirit, Lamentations 3: "My eye has ravaged my soul."[158]

And therefore, as Helinand says: "A chastity that laughs inordinately or gives indecent looks is foolish."[159] Hunters wanting to find and capture wild game usually first send out small hunting dogs called "ferrets" (*fureti*) that do not stop roaming back and forth until they find prey. And when they have located it, the hunter sends out large dogs, and after they have captured the game, the hunter arrives and kills the wild animal with a hunting spear.[160] Likewise, the devil, too, who "hunts

154 Johannes de Hauvilla, *Arch.*, 1.14.404, ed. Schmidt, 141; trans. Wetherbee, 27. The verse is taken from the description of a girl.

155 Cf. Aristotle, *De insomniis*, 3 (459b23–460a23), ed. and trans. David Gallop, in *Aristotle on Sleep and Dreams. A Text and Translation with Introduction, Notes and Glossary* (Warminster, Engl., 1996), 93; cf. Pliny the Younger, *Nat. hist.*, 7.15(13).64 and 28.7(23).82, ed. von Jan and Mayhoff, vol. 2, 23 and vol. 4, 304.

156 Ps.-Quintilian, *Declamationes XIX maiores Quintiliano falso ascriptae*, 1.6, ed. Lennart Håkanson (Stuttgart, 1982), 7.

157 Jer 9:21.

158 Lam 3:51.

159 From Helinand of Froidmont via Vincent of Beauvais. See Vincent of Beauvais, *Spec. hist.*, 17.9, in *Speculum quadruplex*, vol. 4, 656a.

160 Cf. Alfred Tobler, *Tobler-Lommatzsch, altfranzösisches Wörterbuch. Adolf Toblers nachgelassene Materialien*, ed. Erhard Lommatzsch (Berlin, 1954), vol. 3 (q.v. *furet*), described as a small animal used in hunting rabbits. See also du Cange, *Glossarium*, vol. 3 (q.v. *furectus*, *furetus*, and *furo* [2]), all of which refer to an animal used in hunts, especially in hunting rabbits. Peter seems to use the word in an other-

down mankind until death,"[161] sends out first the unchaste eyes of a young man as if they were his ferrets, and they roam through the towns until they locate prey, that is, until they spot some beautiful woman, and then the devil sends to the core of the youth's being the large dogs, that is to say, unclean thoughts, and finally the hunter arrives and kills the soul with the hunting spear of consent or the deed itself. For "whoever looks at a woman to desire her has already committed adultery with her in his heart"[162] and thus has been spiritually killed. For this reason it is called fornication (*fornicatio*) as if it were "the murder of a powerful man" (*fortis necatio*), or "beauty's murder" (*forme necatio*), or "the kindling of the furnace" (*fornacis accensio*).[163] The nature of the lion is said to be such that if someone passes by close to a lion and does not look at it, he escapes unharmed; if he looks at it, however, the lion will attack him in order to devour him. Likewise, through a glance at a woman a man is sometimes unexpectedly attacked by carnal temptation which he would have safely avoided if he had not looked at her. Therefore, it is said in Job 31: "I have sworn an oath with my eyes that I will never look at a girl,"[164] and about this passage Gregory says that: "It is not permitted to look at what it is not permitted to desire."[165] And Isidore in the *Book of Synods*: "Take away the natural propensity for transgression. If you want to be safe from the frightening act (or otherwise: from lust),[166] be discreet in vision."[167] No one with a sane mind would go to look at the basilisk, the sight of which kills. Thus, no one should go to look at women, the sight of whom sometimes murders and kills:

---

wise unrecorded way to refer to small dogs, not ferrets, used to scent and flush out prey. The hunt described here is also not the ceremonious, aristocratic kind described in many hunting manuals, but a more informal type of hunt. For information on medieval hunts, I am indebted to David Scott-Macnab, University of Johannesburg.

161 Mic 7:2.

162 Matt 5:27.

163 For *forme necatio* as the etymology of *fornicatio*, cf. Sedulius Scotus, *Collectaneum in Apostolum, In Epistolam ad Romanos*, cap. 1, v. 29, ed. Hermann Josef Frede and Herbert Stanjek, Vetus Latina. Aus der Geschichte der lateinischen Bibel, 31 (Freiburg i. Br., 1996), vol. 1, 46; Petrus Helias, *Summa super Priscianum*, De genere (Pr. 5), ed. Leo Reilly, Studies and Texts, 113 (Toronto, 1993), vol. 1, 360.

164 Job 31:1.

165 Gregory the Great, *Mor. in Iob*, 21.2.4, ed. Adriaen, CCSL 143A:1066.

166 The reading in *Pr1* (*a formidine alias a fornicacione*) is the result of scribal dittography or an attempt to provide two possible expansions of an abbreviation. *P7P8* reads simply *a fornicacione* at this point.

167 Isidore of Seville, *Synonym.*, 2.16–17, ed. Elfassi, CCSL 111B:74–75 (PL 83:849A–B).

Conversation with them is nothing but bitter venom*.[168]

A certain young hermit went to the city with his abbot, where he saw women at a dance, and being agitated he asked his abbot what kind of thing they were. The abbot responded that these were geese. Having returned to the monastery, the boy began to cry inconsolably. The abbot asked him, "What do you want, son?" And he said, "I want some of those geese that I saw in the city." Then the abbot called together the brothers and said, "Consider carefully, brothers, how dangerous is the sight of women. For this innocent boy, who was raised in the desert and never saw a woman before, has thus been set afire with carnal desire by a single look."[169]

It would be insane for someone with a jar of balsam to set it down in the middle of the square and expose it to everyone so that each and every person could throw a stone at the jar. In the same way, someone who has the balsam of grace in his heart (so to speak, in a fragile jar because of the changeableness of his free will) and who also has the treasure of chastity and blamelessness in his body (which is very easily broken because the body is prone to do something sinful) – that person is insane if he has dared to go look at women and thus expose his jar indiscriminately to stones knocking against it. For the first woman he sees throws a likeness of her beauty, as if it were a stone, at the small jar of his heart. And the same thing happens when he sees the second woman. And however many women he looks at shamelessly are, as it were, so many stones thrown at him. And therefore it is said in Ezekiel 6: "Their lascivious eyes, roving after idols,"[170] i.e., after the likenesses of women that they have perceived; and Seneca in one of his letters: "Nothing is so damaging to good morals as to sit around at some spectacle, for then vices creep in the more easily through pleasure."[171] And Gregory says in a *Dialogue*: "Those who dedicate their hearts to temperance do not dare live with women."[172] For when the beauty of a woman is placed within eyesight, it makes many men have repulsive thoughts.

Just as the sight of a visible object of excessive whiteness is more damaging to the eye than a visible object that is strikingly black, so also

168 Bernard of Morlaix (Bernard of Cluny) (?), *De vanitate mundi* ("Chartula nostra ..."), PL 184:1309B. The reading *virus* in the PL text has been preferred to *virtus* found in *Pr1*.

169 Cf. Tubach, *Index exemplorum*, nos. 1571, 3963, and 5365.

170 Ezek 6:9.

171 Seneca, *Ep. mor.*, 7.2, ed. Reynolds, vol. 1, 12.

172 Gregory the Great, *Dial.*, 3.7.1, trans. Antin, ed. de Vogüé, SC 260:278–79 (PL 77:229A).

the sight of a woman of outstanding beauty is more dangerous than that of an ugly woman, and the sight of one who is decked out in the apparel of a whore is more dangerous than that of one who is clothed modestly. Extravagant clothes give evidence of a woman's lascivious heart and incite onlookers to evil. Hence, there is the poem:

> The outward condition proclaims the spirit; the inner disposition
> Is depicted externally; the attire is a demonstration, affording access
> To a hidden intention; although clothing covers the rest,
> It uncovers the heart and is the garrulous expounder of character,
> Revealing hidden matters; and – with the clothing speaking the
>   mind –
> The outward page announces its inner laws.[173]

Therefore, it is said in Ecclesiasticus 9: "Turn your face from a woman who is made up beautifully, and do not look around for someone else's beauty."[174] For many have perished because of the beauty of a woman. Sigebert relates in his *Chronicles* that a tribune came with his wife to John the Anchorite and asked him to show himself to their sight, for his wife desired this greatly. But John refused to do so, and he added, "Go away, for she'll see me tonight in your home." That night he appeared to the woman and said, "In compliance with your wish, woman, I've come to you, and I caution you to desire to gaze spiritually not upon the bodily appearance of God's servants, but upon their deeds."[175] And it is said in the *Lives of the Fathers* that a certain widow came to the blessed Arsenius from far distant regions and, lying prostrate with tears, she said, "Pray to God on my behalf." He responded to her, "I pray to God that the memory of you will be wiped from my mind," and he turned away from her. Then she groaned and went away.[176] How much more frightened can we weaklings be, therefore, when we hear that the saints avoided the sight of

173 Johannes de Hauvilla, *Arch.*, 2.3.158–64, ed. Schmidt, 149; trans. Wetherbee, 41. The passage is taken from the chapter describing Cupid's clothing.

174 Sir 9:8.

175 Cf. Vincent of Beauvais, *Spec. hist.*, 17.4, in *Speculum quadruplex*, vol. 4, 654b, which is one of the chapters in this text devoted to the life and deeds of John the Anchorite. The previous chapter in the *Speculum historiale* ends with material quoted from Sigebert, which may account for Peter's attribution of the incident related here to Sigebert's *Chronicles*. John ("of Lycopolis"; ca. 310–394/95), was a renowned recluse in the Egyptian desert.

176 *Vit. pat.*, 5.2.7, PL 73:858D–59B. Arsenius ("the Great"; d. 450), from a Roman senatorial family, became one of the most influential Desert Fathers of the early fifth century.

women to such an extent! Because the blessed David cast his eyes on a woman and kept his gaze fixed there, he immediately gave himself up totally to his desire and committed adultery and homicide.[177] For if he had not looked at a woman incautiously, he would hardly have fallen into such a temptation of the flesh. Thus, if such a great man fell through a glance of his eyes, we, who are far from his holiness, ought to beware of and fear a deadly fall to the utmost extent and to pray with this same David and to ask of the Lord: "Turn my eyes away, lest they see untruth."[178]

But a woman captures a man not only by his eyes, but also by each one of his senses, and first of all by the sense of hearing, as exemplified by the Sirens. The Sirens are sea monsters that possess the facial appearance of a maiden and attract sailors by their song, and just as they cast the men they have allured into a deep sleep and kill the slumberers, so also a woman attracts a man with her song and makes him fall asleep in sin, and thus she kills him spiritually.[179] In this deed the one is as guilty as the other: the woman, since she destroys the man through their transgression, and the man, since he approaches at her song like a fool. As fables relate, a certain lady owned a very choice cow which she committed to the care of one of her herdsmen. This herdsman, by name Argus, is reported to have had a hundred eyes, according to the verse:

> Argus' head was circled round by a hundredfold eyes.[180]

Wanting to guard the cow with the utmost diligence, Argus allowed only one pair of eyes to sleep at any one time, and in the meantime the other ninety-eight were awake and keeping watch, guarding the cow. But someone else who was desirous of owning the woman's cow sent his son, named Mercury, to Argus with a musical instrument. After he had arrived, Mercury began to recite fables to Argus in the manner of an actor and to sing quite a number of them, and he did not stop doing this until two of Argus' eyes had begun to doze off in sleep. Then two others fell asleep to Mercury's song, and so it continued until all of them had been seized by sleep. When Mercury saw this, he immediately cut off Argus' head and took the cow and brought it to his father and reported to him how he got the better of Argus through the allurement of song.[181] This Argus, who was furnished with a hundred eyes, stands for a man who is

177 Cf. 2Sam 11:2–15.

178 Ps 118:37.

179 Cf. Tubach, *Index exemplorum*, no. 4495.

180 Ovid, *Met.*, 1.625, ed. Anderson, 21 (cf. Walther, *Initia carminum*, no. 2613).

181 For the story of Argus and Mercury, cf. Tubach, *Index exemplorum*, no. 1298.

prudent and circumspect in all ways, but nevertheless if he willingly listens to a woman's song, it must be feared that he will falter in his circumspection and vigilance and perish spiritually while his precaution sleeps, and through sin lose his soul which he ought to have protected carefully. Hence, Jerome says in one of his letters: "It is a matter of great virtue and very careful diligence not to live carnally although you were born in the flesh, to fight against yourself daily, and to keep watch against the enemy imprisoned inside you with a hundred eyes, as fables relate of Argus."[182]

Second, a woman captures through the sense of sight, as exemplified by the tiger. For when a tiger perceives a mirror placed in its path intentionally by hunters, it halts in its tracks, charmed by its own beauty in the mirror, and forgets its cubs.[183] In the place of the bright mirror, understand the adorned body of a woman; when someone who is a fool looks at it, it sometimes transfixes him in his tracks spiritually and leads him to forget the heavenly things which he should pursue.

Third, by the sense of smell, as exemplified by the panther. As the *Physiologus* says, a panther uses the fragrance of its odor to attract other animals, and after they have become aware of its sweet odor they will accompany the panther even to their deaths.[184] And in the same way, foolish men, whose hearts are charmed by vain odors, very often run after the scent of perfumed women. But let women who devote their attention to such things hear what the Lord threatens through Isaiah in Isaiah 3: "Because the daughters of Zion are haughty and walk with outstretched necks and give beckoning glances with their eyes," and so on.[185] And this passage follows: "On that day the Lord will take away their scent bottles and ear-rings and mirrors, and instead of a sweet smell there will be rottenness."[186]

Fourth, a woman captures a man by the sense of touch, that is to say, by performing the abominable deed, and then the unfortunate man is corrupted, on the example of the unicorn. As those people claim who have written about the nature of animals, the unicorn is captured in this way: Namely, a young virgin is set out in the forest becomingly attired, and she makes room on her lap when the unicorn approaches. It lays its

182 Jerome, *Ep.*, 54.9.3, ed. Hilberg, CSEL 54:475.

183 Cf. Tubach, *Index exemplorum*, no. 4865.

184 Cf. *Physiologus*, 2.2–3 (xxiv), ed. Friedrich Maurer, in *Der altdeutsche Physiologus: Die Millstätter Reimfassung und die Wiener Prosa (nebst dem lateinischen Text und dem althochdeutschen Physiologus)*, Altdeutsche Textbibliothek, 67 (Tübingen, 1967), 75.

185 Isa 3:16.

186 Isa 3:18–24.

head in her lap, having rid itself of all ferocity, and after it has fallen asleep, the hunters capture and kill it.[187] In the same way, Samson lost his eyes sleeping on Delilah's lap, as a result of which he eventually incurred death, Judges 16.[188] For this reason Gregory says in the *Morals*: "In accordance with the exacting decree of retribution it happens that whoever uses his exterior eye negligently is blinded not unjustly in his interior one."[189] Eusebius also says in the *Chronicles* that because of her extraordinary beauty the whore Gorgona made those looking at her so weak-minded that one would think they had been turned into stone.[190]

When the raven finds a dead animal, it first pecks out the eyes for itself and then extracts the brain.[191] So also a woman first captures a man through the gaze of his eyes, and then draws his brain to her when she takes his senses from him and turns him into a simpleton. For:

> Every lover is blind, love is not an impartial observer;
> For it judges a misshapen cow to be a thing of beauty.[192]

As Solinus says in the *Wonders of the World*, this is the way to capture apes, for since they have a natural disposition to mimic others, they eagerly try to reproduce the gestures of the hunters by smearing their eyes with the birdlime left in the forest by the hunters, just as they saw the hunters pretend to have done. And when their eyes have been closed tight, it is easy to capture them.[193] So, too, a shameless woman beautifies herself

187 Cf. Isidore of Seville, *Etym.*, 12.2.13, ed. Lindsay, vol. 2 (cf. Tubach, *Index exemplorum*, no. 5023).

188 Cf. Judg 16:19–21.

189 Gregory the Great, *Mor. in Iob*, 21.8.13, ed. Adriaen, CCSL 143A:1075.

190 Eusebius of Caesarea, *Chronicle*, trans. Jerome (Saint), anno 670 (from Abraham), ed. Rudolf Helm, in *Eusebius Werke 7: Die Chronik des Hieronymus*, Die griechischen christlichen Schriftsteller der ersten drei Jahrhunderte, 47, 2nd ed. (Leipzig, 1956; reprint Berlin, 1984), 81 (in Jerome) (cf. Tubach, *Index exemplorum*, no. 3255).

191 Cf. Hugh of Fouilloy (Ps.-Hugh of Saint-Victor), *Aviarium*, 2.40, ed. and trans. Clark, 174–75. Cf. also Ps.-Hugh of Saint-Victor, *De bestiis et aliis rebus. Libri iii–iv*, 4.3, PL 177:143A. Books 3 and 4 of this text, erroneously printed under the name of Hugh of Saint-Victor, were heavily edited and reordered by their sixteenth-century editors; on the text, see most recently Willene B. Clark, "Commentary," in *A Medieval Book of Beasts. The Second-Family Bestiary*, ed. and trans. Willene B. Clark (Woodbridge, 2006), 7–116, here 13–14.

192 *Sententia communis* (Walther, *Proverbia*, no. 20188).

193 Cf. Gaius Julius Solinus, *Collect. rer. mem.*, 27.56, ed. Mommsen, 127. The *exemplum* is discussed by Janson, *Apes and Ape Lore*, 15, 34.

when she makes up her eyes in order to be captured by men. For this reason, while she makes her external face radiant, she conceals her internal one, Jeremiah 4: "Though you paint your eyes with make-up, you beautify yourself in vain."[194] After a certain whore had lost one eye due to a disease and a priest told her this had happened through God's fitting vengeance because of her sin, she said, "I prefer to be content with one eye rather than with one man."[195] Behold a woman deprived of her eyes because of her incontinence – she lost one from her body, but two from her mind! This blindness of carnal desire is represented in Saul's blindness, Acts 9.[196] For he was following the religious rites of the law which taught carnal observances, but he was cured by falling down on the earth, that is, by tormenting his flesh, and by showing himself to Ananias, that is, to a priest in confession.

This blindness that originates in the sin of lust is also depicted with its remedy in the blindness of Tobit. Indeed, one reads that Tobit, blinded by the excrement of swallows, was cured by the gall bladder of a fish placed over his eyes.[197] In the excrement, lust is signified, Ecclesiasticus 9: "A woman who is a fornicator shall be trodden on like excrement on the road by men who pass by."[198] The fish, which is nourished in the water, stands for the sin of the flesh, which is nourished in the flow of temporal riches, since:

> Poverty owns nothing from which it feeds its passion.[199]

A fish has sweet-tasting flesh, but its gall bladder is bitter; carnal sin is also sweet during the act, but the feeling of guilt it causes is bitter. Thus, if the bitterness of the gall bladder, not the delightfulness of bodies, is placed over the eyes of the mind, that is, if a man considers and pays attention not to the sweetness of carnal delight, but to the bitterness of the punishment to come, he will be cured of the blindness of carnal desire by this kind of medicine.

194 Jer 4:30.

195 For the conception of a woman being satisfied with one eye rather than limited to having only one man, cf. Juvenal, *Sat.*, 6.53–54, ed. Clausen, 73.

196 Cf. Acts 9:8–9.

197 Cf. Tob 2:11, cf. Tob 6:5.

198 Sir 9:10.

199 Ovid, *Rem. am.*, 749, ed. Kenney, 258 (cf. Walther, *Proverbia*, no. 17844).

CHAPTER NINE

On the Costliness of the Eyes in Respect to the Other Members of the Human Body

At the table of a certain prelate this question was raised: Which part of the human body leads to more expenses, generally speaking? And everyone responded that the stomach is the part that leads to more expenses, according to the verse in Ecclesiastes 6: "All of a person's toil is for his mouth."[1] But finally, when a certain philosopher who was sitting at the table with the others was asked what he felt about this matter, he said, "I think the eyes are more costly than any other part of the human body, since there are six things that eat up excessive expenses from the rich, namely the costliness of clothes, horses, buildings, books*, food, vessels of silver and also those of gold." For superfluity of this kind has been dreamt up for no other reason than to feed the eyes of the rich and, what is even more deplorable, others' eyes as well. For this reason Seneca says: "It is not hunger which costs us so much, but eager cravings."[2] And in the *Book of Morals*, the same author says: "No longer are people content with plundering by their teeth and stomach and mouth; they are also gluttonous with their eyes."[3] And someone has said:

> The eyes feed the heart more than dishes do the stomach;
> The honor of others' obsequiousness tastes better than a good
> flavor itself.[4]

And for this reason the poet says when describing the avidity of the gluttonous man's eyes:

> Thus he looks over the feast, and with his eyes and his will
> Already devours the provisions that have been laid out; he examines
> all

1 Qoh 6:7.

2 Seneca, *Ep. mor.*, 60.3, ed. Reynolds, vol. 1, 167.

3 Seneca, *Naturalium quaestionum libri*, 3.18.7, ed. Harry M. Hine (Stuttgart, 1996), 134.

4 *Sententia communis* (Walther, *Proverbia*, no. 21733).

The opulence on the table, and he measures it once more:
He inspects and looks over the platters with his caressing eyes.[5]

And in this, the miraculous power of the eye becomes apparent, since it can even change gold into food. For while gluttony is nourished by food, the eye is nourished by gold and silver. And so the poet says:

Metal is made into a delicate meal.
Its inborn hardness is transformed to be used at a banquet,
And gold is compelled by princes to counterfeit* food.[6]

Helinand relates that in the year of the Lord 1092* when a certain king of England named William put on new leather shoes one morning, he asked his chamber-servant what price they had been bought for. And when the man responded, "For three shillings," the king became indignant and enraged and said, "You son of a bitch, why does the king own shoes of such a meager price? Go, in the name of the devil, and bring me shoes for a mark of silver!" The man went out and brought back shoes which were much cheaper*, but he pretended that they had been purchased for as much as the king had ordered. "These," said the king, "on which the eyes of men must be focused, are fit for royal grandeur!"[7] In order to gratify others' eyes, this man transformed pennies into shillings. Such is the costliness of the eyes that Solomon says of himself in Ecclesiastes 2, "Everything which my eyes desired I did not keep from them,"[8] and he also says elsewhere, "The eyes of human beings are insatiable."[9] And Ecclesiasticus 14: "The eye of the greedy person is insatiable."[10]

Moreover, the excessive curiosity of the eyes is such that it not only causes people to go beyond all measure in expenses, but it also sometimes causes them to go insane, according to the passage in Ezekiel 23: "The desire of her eyes drove her insane."[11] We can see this in Nero, for

5 Johannes de Hauvilla, *Arch.*, 2.12.362–64 (first half of the line), 2.12.350 (first half of the line), ed. Schmidt, 156; trans. Wetherbee, 51, 53. The verses, excluding the last one, are taken from the description of the glutton's condition while eating.

6 Ibid., 2.7.250–52, p. 152; trans. Wetherbee, 45. The passage is taken from a brief chapter against those who turn gold into food.

7 From Helinand of Froidmont, *Chronicon*, 47.anno 1092, PL 212:981A–B, probably via Vincent of Beauvais, *Spec. hist.*, 25.85, in *Speculum quadruplex*, vol. 4, 1032a (cf. Tubach, *Index exemplorum*, no. 1119). The reference is to William II ("Rufus"; ca. 1056–1100), king of England from 1087.

8 Qoh 2:10.

9 Prov 27:20.

10 Sir 14:9.

11 Ezek 23:16.

example. For as Eusebius bears witness in his *Chronicles* and Boethius in the *Consolation of Philosophy*, in order to satisfy the unbridled hunger of his eyes, Nero set fire to the city of Rome so that he could see the likeness of Troy burning.[12] For this reason Orosius says: Wanting to see an imitation of the Trojan fire, the emperor Nero ordered the burning of Rome as a spectacle for his pleasure. The burning city fed his regal sight for six days and seven nights. Indeed, gazing over the conflagration of the city from the highest tower, he said he was delighted by the beauty of the flames, and he sang the *Iliad*[13] in the costume of a tragic actor. And, what is even more offensive to hear, he ordered that his mother be disemboweled while she was still alive so that in this way he could see the place in which he had lain for a long time before birth.[14] Clearly, this man's eyes had made him demented, for they enticed him into perpetrating such heinous deeds. Because of this, not inappropriately it is asked in Ecclesiasticus 31: "What has been created more evil than the eye?"[15] Not only the attestations of Holy Scriptures but also the fables of the poets give evidence of how pernicious an excessive curiosity of the eyes can be. For this reason Fulgentius says as follows in book 3* of the *Mythologies*: An excessive curiosity of the eyes always knows how to bring about harm to those charmed by it, a harm that is closely related to the dangers it leads to. In this way, the hunter Actaeon was motivated by an excessive curiosity of the eyes to want to see Diana bathing, and one reads that he was changed into a stag and was devoured by his own dogs, which did not recognize him.[16]

12 Cf. Eusebius of Caesarea, *Chron.*, trans. Jerome, anno 2080 (from Abraham), ed. Helm, 265 (in Jerome); cf. Boethius, *Phil. cons.*, 2.metr.6.14–17, ed. Bieler, CCSL 94:31–32.

13 *Pr1* has *iliadam* here; in *P7P8* this is *Heleifeles* (*P7*: *Heleipheles*) *odam*, i.e., through corruption, the *Halosis Ilii* (an ode on the defeat of Troy); cf. Suetonius, *De uita Caesarum libri viii*, Nero, 38.2, ed. Maximilian Ihm, ed. min. (Stuttgart, 1908; reprint 1967), 248 (cf. Petronius, *Satyricon*, 89.1, ed. Mueller, 88).

14 Cf. Paulus Orosius, *Historiarum adversum paganos libri vii*, 7.7.4, 6, ed. Karl Zangemeister, CSEL 5 (Vienna, 1882), 453–54.

15 Sir 31:15.

16 Cf. Fulgentius the Mythographer, *Mitologiarum libri tres*, 3.3, ed. Rudolf Helm and Jean Préaux, in *Opera* (Stuttgart, 1970), 62–63; cf. Ovid, *Met.*, 3.143–252, ed. Anderson, 58–62.

CHAPTER TEN

On Suffering the Loss of the Corporeal Eyes with Equanimity

After Tobit became blind, one reads that he gave thanks in return to God, Tobit 2.[1] For sometimes the loss of the eyes works out for the best for someone. Seneca says in *On the Remedies of Fortune*: "One ought to understand that to have lost one's eyes is to have gained a portion of innocence. To one person the eyes reveal adultery; to another, incest; to yet another, a house he yearns to have. For the eyes are incitements to vices and the instigators of evil deeds."[2] Not to have had eyes is better than to have offended God with one's eyes. For this reason Cicero says in his *Orations*: "When you have seized on every* act of deceit with your eyes, you are more pitiable than if you had not had eyes altogether."[3] Hence also, as the *Policraticus** asserts in book 8: "One reads that some philosophers tore out their eyes so they would not be captive to external allurements. They were undoubtedly aglow with an admirable zeal for what is good and honorable, although they had no knowledge of what is right, since indeed, as Aristotle says, the beginning steps of pleasure ought to be consigned to oblivion, and only the ways in which pleasure ends filled with repentance ought to be remembered."[4]

And therefore when those who are holy consider diligently the evils arising from the eyes, they glory in the deprivation of one of their eyes or, if it should happen, of both of them. For this reason one reads that when Petrus Monoculus, abbot of Clairvaux, had lost one of his eyes due to a disease, this man of God used to say as if in jest that he had escaped from one of his enemies and he was more frightened by the remaining one than by the one he had lost.[5] In the same way after another holy man had

1 Cf. Tob 2:14.

2 Ps.-Seneca, *Ad Gall. de remed. fortuit.*, 12.1–2, ed. Haase, 454.

3 Cicero, *Oratio de haruspicum responsis*, 18.39, ed. William Peterson, in *Orationes* (Oxford, 1910; reprint 1962), vol. 5.

4 John of Salisbury, *Policr.*, 8.6, ed. Webb, vol. 2, 251 (cf. Tubach, *Index exemplorum*, no. 1945); for Aristotle, see Valerius Maximus, *Fact. et dict. mem.*, 7.2.Ext.11, ed. Briscoe, vol. 2, 449.

5 Cf. Thomas de Radolio, *Vita Petri abbatis Claraevallis*, 26, PL 209:1029C. Petrus Monoculus was abbot of Clairvaux, 1179–1186.

lost one eye, he said to his friends who were weeping, "Don't weep, for of two enemies, I've lost one."[6] One reads in book 6 of the *Tripartite History* that when Julian the Apostate was sacrificing to idols in Constantinople, Maris, bishop of Chalcedon, came to him and publicly called him an apostate. But Julian upbraided him only for his blindness, for he was an old man and was guided by someone else's hand because of the weakness of his eyes. And the ungodly Julian added, "Nor can your man from Galilee cure you." "I give thanks to my God for this blindness," said Maris, "for it has happened so that I don't have to see you, devoid of all piety."[7]

Further, Sigebert relates in the *Chronicles* and the same thing is found in the *Life of Saint Audomar*: Audomar, bishop of Thérouanne*, became blind due to old age and after he had regained his sight through divine providence at Arras during the ceremonial translation of Saint Vedast, he grew sorrowful that he had lost the blindness he had counted as a gift from God, and his prayers were answered that he become blind again. And thus externally blind, he was internally enlightened, saying, "I have raised my eyes to you who dwells in heaven."[8] For it often happens that when the light of the corporeal eyes is lost, that of the spiritual eyes is rejuvenated. One reads in book 8 of the *Tripartite History* that Didymus, blind from childhood, learned through hearing alone and knew by heart grammar, logic, rhetoric, arithmetic, geometry, and the other philosophical disciplines. Beyond that, he knew the New and Old Testament with such perfection that he published many books on them and wrote three volumes on the Trinity and he left us excellent interpretations of Origen's book, *On Principles*. When Antony saw him, he said to him, "Don't let the loss of your sensory eyes disturb you, Didymus, for you are without the eyes that flies and fleas can bother, but rather rejoice since you possess the eyes that give angels sight as well, and with them God is seen and his light is examined."[9]

6 Cf. Tubach, *Index exemplorum*, no. 1939.

7 Cassiodorus-Epiphanius, *Historia ecclesiastica tripartita*, 6.6.1–2, ed. Walter Jacob and Rudolf Hanslik, CSEL 71 (Vienna, 1952), 317. Maris was an Arian bishop and opponent of Athanasius who flourished around the middle of the fourth century.

8 Ps 122:1. Cf. Sigebert of Gembloux, *Chronica*, anno 658, ed. Ludwig Bethmann, MGH, Scriptores, 6 (Hannover, 1844), 326; cf. *Vita s. Audomari Episcopi Terwennensis*, 11, PL 147:1186B–C. Audomar (d. ca. 670) was made bishop of Thérouanne in 637.

9 Cassiodorus-Epiphanius, *Hist. eccl. tripartita*, 8.8.1–6, ed. Jacob and Hanslik, CSEL 71:478–80. Didymus (d. ca. 395/98) was fabled for his learning, in spite of being blind, and was the head of the Alexandrian catechetical school. Saint Antony (d. middle of the fourth century) was a formative figure in the development of anchoritic monasticism.

Hence, "we also read that some heathen philosophers tore out their eyes in order to turn all of their thoughts towards a purity of mind," as Jerome says, *Against Jovinian*.[10] And Claudius* testifies in his book that the philosopher Democritus did this.[11] And therefore Aulus Gellius in the *Attic* Nights* and Justinus, as well, say that the "disciples of Plato passed their time in the groves and arcades of temples so that the holiness of a very restrained dwelling would warn them to think of nothing other than virtues." Because of this desire some of them are also said "to have torn out their eyes lest they be drawn away from their contemplation of wisdom by vision with the eyes."[12]

10 Jerome, *Ep.*, 76.2.2, ed. Hilberg, CSEL 55:35.

11 Cf. Aulus Gellius, *Noctes Atticae*, 10.17.1, ed. Peter K. Marshall (Oxford, 1990), vol. 1, 321. Gellius quotes often from the *Annals* by the first-century (BCE) Roman historian, Quintus Claudius Quadrigarius, referring to him at times simply as Q. Claudius. This is apparently the source of the reference to Claudius here.

12 Jerome, *Adv. Iov.*, 2.9, PL 23:298B–C. The first sentence of this passage is also found verbatim in John of Salisbury, *Policr.*, 8.8, ed. Webb, vol. 2, 278–79. I have not found a statement like this in the *Attic Nights* or the works of Justin Martyr.

CHAPTER ELEVEN

On the Education of Students Based on the Seven Conditions That Are Required for Sight

For vision to be completed, these seven things are necessary: the required disposition of the visual organ, the presence of an object, a proportionate amount of distance, the density of the visual object, the application of the visual faculty, a duration of time, and the diffusion of light illuminating the medium.[1]

The first lesson. On humility in learning, being scornful of no one, and enduring insults.*

These three things obliterate the required disposition of the eye: swelling, dust, and fluid. Thus, spiritually, these are: the swelling or elevation of pride, the dust of avarice, and the congealed fluid of lust. About the first one Aristotle says in book 19 of *On Animals* that an eye which protrudes is weak of sight; an eye which is deep set has strong and good

1 Cf. Roger Bacon, *Persp.*, 1.8.1–9.4, ed. and trans. Lindberg, 108–44, who gives nine requirements for vision: the visual *species*, light, distance between the eye and the object it perceives, the location of the visible object opposite the eye, the perceptible size of the visible object, the density of the visible object, the rarity of the medium (i.e., no dense object can be placed between the eye and the visible object), a passage of time, and a healthy eye possessed of its natural disposition. Cf. also Alhacen, *De aspect.*, 1.9 [8.1], ed. and trans. Smith, in *Alhacen's Theory of Visual Perception*, vol. 1, 72–73 (Latin text), vol. 2, 390–91 (English translation), who maintains that six conditions must be met for vision to occur: distance between the eye and the object it perceives, the location of the visible object opposite the eye, the illumination of the visible object, the perceptible size of the visible object, the transparency of the medium between the eye and the visible object, and the density of the visible object. In *De aspect.*, 3.3 [3.5], vol. 1, 286 (Latin text), vol. 2, 588–89 (English translation), Alhacen expands this list of the conditions to be met for the perception of an object as it actually exists, including two more factors: the passage of time and a healthy eye. It should be noted that Peter's fifth condition (the application of the visual faculty) is nowhere stated explicitly by either Bacon or Alhacen.

sight.[2] In a similar fashion, the same can be said about a humble person. For people with deep-set eyes see things from far off in the distance, and also intricate and minute things, better than people with swollen or protruding eyes, which are also easily injured by the sensible object: Likewise, humble people see intricate matters more clearly than the proud, Proverbs 11: "Where there is humility, there is wisdom."[3] Therefore, the Savior said in Matthew 11, "I will confess to you, father, Lord of heaven and earth, since you have hidden these things from the wise and prudent and have revealed them to children,"[4] that is to say, "to the humble."[5] For sometimes the Lord reveals things to children which he conceals from the wise. As an indication of this matter, in Samaria salvation was announced through lepers,[6] and Balaam's ass, though accustomed to the yoke, turned its master aside from the path which had been forbidden.[7] And for this reason the hermit Antony asked a number of fathers about the Scriptures and everyone responded according to what he knew. But Joseph, whom he questioned last, said, "I don't know." And Antony said, "Only this man has found the path to knowledge who responded that he doesn't know."[8] Consequently, Bernard says in *The Steps of Humility*: "Likewise, that the Lord appeared above the ladder which was shown* to Jacob as a figure of humility, what does this intimate to us if not that the perception of truth is founded upon the summit of humility?"[9] And Isidore in book 2 of the *Soliloquies*: "Truth flees from a mind that has not discovered humility."[10] And concerning Socrates, Valerius Maximus says in book 8 that while he considered himself a pauper when it came to

2 Cf. Aristotle, *De gen. animal.*, 5.1, trans. William of Moerbeke, ed. Drossaart Lulofs, 160; and trans. Michael Scot, ed. van Oppenraaij, 220; idem, *De hist. animal.*, 1.10, trans. William of Moerbeke, ed. Beullens and Bossier, 18–19. The same passage from Aristotle on the visual acuity of deep-set eyes is referred to by Roger Bacon, *Persp.*, 2.1.1, ed. and trans. Lindberg, 162.

3 Prov 11:2.

4 Matt 11:25.

5 *Gloss. ord.*, Matt. 11:25, ed. Rusch, vol. 4, 42b interlinear.

6 Cf. Luke 17:11–19.

7 Cf. Num 22:23.

8 *Vit. pat.*, 5.15.4, PL 73:953C. This is Saint Antony (d. middle of the fourth century), a formative figure in the development of anchoritic monasticism. The Joseph is perhaps Joseph of Panephysis, to whom several sayings are attributed in the alphabetical collection of the sayings of the Desert Fathers.

9 Bernard of Clairvaux, *Liber de gradibus humilitatis et superbiae*, 2.3, ed. Jean Leclercq and Henri M. Rochais, in *Opera* (Rome, 1963), vol. 3, 18; cf. Gen 28:13.

10 Cf. Defensor of Ligugé (?), *Liber scintillarum*, 4.31, ed. Henri M. Rochais, CCSL 117 (Turnhout, 1957), 20; cf. Ps.-Augustine of Hippo, *Sermones*, 297.1, PL 39:2314.

learning, in matters concerned with teaching he made himself the wealthiest person.[11]

Thus, in order for a student to make good progress in knowledge, it is necessary for him to strive for humility, since according to the Psalm, the Lord "gives understanding to children,"[12] i.e., to the humble. Revelation 3: "Behold, I have set before you an open gate," that is, the entrance to the knowledge of Scriptures, "since you have only moderate (*modicam*) powers,"[13] explained in the *Gloss* as "humility, through which a person is considered ordinary (*modicum*) in his own eyes";[14] and Deuteronomy 28: "If you keep the commandments of your God,"[15] "the Lord will open his good treasure, the heavens,"[16] i.e., the knowledge of heavenly matters. The Lord's commandment is humility, Matthew 11: "Learn from me, since I am gentle and humble of heart."[17]

The humility of a student consists in three things, the first of which is that he should learn willingly from anyone. For this reason Augustine says: "I, an old man, a bishop of so many years, I am prepared to be taught by even the least of my colleagues."[18] And in this way, Paul, gifted in the wisdom of the law (since he had been taught by Gamaliel),[19] was sent to Ananias, a simple man, for instruction.[20] For this reason it is said to Paul in Acts 9: "Arise, and go into the city, and you will be told what you must do there,"[21] where the *Gloss* says: "An example of learning is given here, and pride is censured, so that no one will think someone else unworthy who could teach* him what he does not know."[22]

The next thing is that the student should not scorn any knowledge. For in the *Similitudes* Anselm says: "Foolish paupers closely resemble dull-witted students, that is to say dunces, for when an unwise pauper gets a small coin or any other type of little present, he says, 'Why should I keep this trifle? I won't be rich because of it; and so I'll spend it on fruit or nuts or whatever else delights me, because I don't want to be bothered

11 Cf. Valerius Maximus, *Fact. et dict. mem.*, 8.7.Ext.8, ed. Briscoe, vol. 2, 527.
12 Ps 118:130.
13 Rev 3:8.
14 Cf. *Gloss. ord.*, Apoc. 3:8, ed. Rusch, vol. 4, 553b interlinear (PL 114:716C).
15 Deut 28:9.
16 Deut 28:12.
17 Matt 11:29.
18 Augustine of Hippo, *Epistulae*, 75(250).2, ed. Alois Goldbacher, CSEL 57 (Vienna and Leipzig, 1911), 595.
19 Cf. Acts 22:3.
20 Cf. Acts 9:17–19; cf. Acts 22:12–16.
21 Acts 9:7.
22 *Gloss. ord.*, Act. 9:7, ed. Rusch, vol. 4, 474b in margin (PL 114:448B).

safeguarding it? In the same way, while the dim-witted student disdains things of the smallest value, he never advances to things of greater worth. But the wise pauper carefully safeguards the few things he acquires, and the less he thinks he possesses, the more he wants to retain those things, so that eventually from a mass of very small things he will be able to attain greater ones. Whoever considers himself a pauper in knowledge ought to act in the same way if he wants to achieve any progress in knowledge. For just as many grains of flour go to make up dough and many drops a river and many coins a rich man, so also many bits of knowledge add up to wisdom."[23]

Next, patience is required so that a student may patiently endure the injuries and affronts that afflict him. About this matter one reads that the ancient philosophers were most zealous, and for this reason a certain philosopher said to a person who was cursing him, "You've learned to curse, and I've learned to have contempt for curses."[24] One also reads that once when Socrates was passing through the public square and someone hit him in the head with his fist, Socrates responded only by saying to the man who hit him, as if in jest, "It's bad when people don't know when they ought to walk around armed and when unarmed. For if I'd known about this punch beforehand, I would've covered my head."[25] Concerning the same philosopher one reads that when he wanted a wife, he married the most contentious woman he could find. As soon as he entered the house, she used to immediately accost him with grievances and accusations. But he, as the most patient of men, kept silent about everything. And when he was asked by his friends why he had married such an annoying woman, he answered, "So that I learn at home what I should be like in the marketplace."[26] Speaking of the patience of this same Socrates, Jerome, too, says in the book *Against Jovinian*: "It happened once that Socrates stood patiently for a long time while his wife was on the balcony of their house showering him with endless insults when she finally drenched him with dirty water. But he responded with nothing more than wiping off his head and saying, 'I knew that after such thundering the rain would follow.'"[27] Once, when the study of philoso-

23 Ps.-Anselm of Canterbury, *Lib. de simil.*, 156, PL 159:689A–C (cf. Alexander of Canterbury, *Lib. ex dict. b. Anselmi*, 3, ed. Southern and Schmitt, 120).

24 John of Salisbury, *Policr.*, 3.14, in *Policraticus I–IV*, ed. Keats-Rohan, CCCM 118:222 (of Xenophon).

25 Cf. Seneca, *De ira*, 3.11.2, ed. Reynolds, 103. The passage in Seneca is quoted by Roger Bacon, *Opus majus*, 7.3.8, ed. Bridges, vol. 2, 291.

26 Cf. the closely contemporary narration of this anecdote in Iacopo da Varazze (Jacobus de Voragine), *Chron. civ. Ianuensis*, 9.1, ed. Monleone, vol. 2, 189–90.

27 Jerome, *Adv. Iov.*, 1.48, PL 23:291B–C.

phy was flourishing at Athens, a certain old man sat at the Portico of the city and when any student came to study, this old man would provoke him with insults, and if he saw in the student a sign of impatience, he would not permit him to enter, but rather forced him to go away as someone unfit for study.[28]

Thus, the student's internal eye should not be conspicuous and swollen with pride, but deep set with humility. Moreover, Perspectivist authors normally explain why people with protruding eyes see worse than those having deep-set eyes by referring to a threefold cause. The first reason is that a deep-set eye is more distanced from external injuries. The second is that the visual power has more support in a deep-set eye, since the eye is more deeply buried. The third reason is that a deep-set eye lies nearer to the common nerve, and the source of the visual power is located in this nerve.[29]

Now, for similar reasons wisdom is granted to the humble rather than to the proud. First of all, since humility is especially distanced from dangers and consequently knows how to avoid dangers. And for this reason when the blessed Antony saw that the world was full of traps and he asked who could avoid these traps, the answer was, "Humility."[30] Nor is it surprising if humility avoids these traps, since if they are on high, humility avoids them since its head bends downward*; if they are underfoot, humility avoids them since it looks at the earth. This, then, is the origin of wisdom: to know how to avoid spiritual evils. Second, since humility does not make a show of its own good qualities, but rather gathers them together and hides them, and the more it does so, the more perfectly it sees. For this reason, as Pliny says in book 2 of *Natural History*: "An eye located in a deep well is able to see the stars in the heavens during broad daylight."[31] Therefore, Ptolemy says at the beginning of the *Almagest*: "Among the wise, whoever is more

28 Cf. Tubach, *Index exemplorum*, no. 3622. The "Portico" indicates the school of the Stoics.

29 Roger Bacon, *Persp.*, 2.1.1, ed. and trans. Lindberg, 160, gives the same three reasons for the visual acuity of deep-set eyes, though in a different order: proximity to the common nerve, protection against injury, and the concentration of the visual power. The concentration of visual power occurs in a deep-set eye, according to Bacon, because the narrowness of the eye socket constricts the visual power in a way analogous to someone looking at the heavens from the bottom of a well, as cited below from Pliny by both Bacon and Peter of Limoges.

30 *Vit. pat.*, 3.129, PL 73:785B–C; ibid., 5.15.3, PL 73:953B–C (cf. Odo of Cheriton, *Parabolae*, 142, ed. Léopold Hervieux, in *Les fabulistes latins depuis le siècle d'Auguste jusqu'à la fin du moyen âge*, 2nd ed. [Paris, 1896], vol. 4, 320; cf. Tubach, *Index exemplorum*, no. 274).

31 Pliny the Younger, *Nat. hist.*, 2.14(11).58, ed. von Jan and Mayhoff, vol. 1, 145. The passage is referred to also in Roger Bacon, *Persp.*, 2.1.1, ed. and trans. Lindberg, 162.

humble has more wisdom, just as a deeper lake is filled with more water than other lakes."[32] Third, since humility is especially close to God. For this reason one reads that the Blessed Virgin, who was most humble and because of this was close to God on high, kept the Savior's every word in her heart,[33] and therefore she is believed to have received a fuller knowledge of divine matters than others, Deuteronomy 33: "Those who approach his feet will receive his instruction."[34] Thus, whoever wants to attain true instruction must come through humility to him who "is the Lord of knowledge," 1 Samuel 2,[35] "in whom all the treasures of wisdom and knowledge have been hidden," Colossians 2.[36] For whoever wants to rejoice in the treasure must be a friend of the treasurer. One must, indeed, humbly knock on his door whose hand holds the book of everything that should be known, which only "the lamb who was slain" "is worthy to open," Revelation 5.[37] He is the key which "opens and no one shuts, shuts and no one opens,"[38] and without which no one attains a true understanding. For he is the one who laid open the meaning for his disciples so they would understand Scriptures when their "eyes were kept from recognizing"[39] the wisdom which he was addressing to them.[40] For whoever presumptuously, through the powers of his own wits "or his study, attempts to break down the sanctity of the Scriptures as if it were a door becomes an outsider excluded from the understanding of the truth as if from the sacristy of wisdom. For also when the men of Sodom wanted to corrupt the angels who had reproved their wickedness and were the guests of a just man, they were struck with blindness, walking around into the walls like blind men

32 This is found among the proverbs attributed to Ptolemy in the preface to Gerard of Cremona's Latin translation of the Arabic version of the *Almagest*: Gerard of Cremona (?), *Praefatio in Almagestum*, 9, ed. and trans. Charles Burnett, "'Ptolemaeus in Almagesto dixit': The Transformation of Ptolemy's *Almagest* in Its Transmission via Arabic into Latin," in *Transformationen antiker Wissenschaften*, ed. Georg Toepfer and Hartmut Böhme (Berlin and New York, 2010), 115–40, here 126 (trans., 129). The prologue accompanied Gerard's translation of the *Almagest* from its earliest transmission, but Burnett (116 and n6) notes that it might not be the work of Gerard himself. For the earliest print of the prologue, see *Praefatio*, in *Almagestum Cl. Ptolemei Pheludiensis Alexandrini Astronomorum principis: Opus ingens ac nobile omnes Celorum motus continens. Felicibus Astris eat in lucem: Ductu Petri Liechtenstein Coloniensis Germani* (Venice, 1515), 1r.

33 Cf. Luke 2:19.

34 Deut 33:3.

35 1Sam 2:3.

36 Col 2:3.

37 Rev 5:12; Rev 5:9.

38 Rev 3:7.

39 Luke 24:16.

40 Cf. Luke 24:27.

and people enveloped by darkness.[41] Thus, while the haughty walk into the wall of Scriptures, they become blind by God's just judgment, and they do not recognize the correct meaning of Scriptures, that is to say, the angel who is the reprover of wickedness and who is in the home of simplicity with whom God is in discussion.[42] Indeed, the Holy Scriptures are like an angel sent by God to rebuke the wickedness of human beings. The angel descends into Sodom and Gomorrah and says, 'Hear God's word, oh princes of Sodom! Give ear to the law of your* God, oh people of Gomorrah,'"[43] Isaiah, the first chapter.[44]

On the love of poverty among students*.

The second thing which hinders the required disposition of the eye is the dust of avarice. And for this reason philosophers formerly cursed riches, nor did they feel that anyone who was anxious about wealth could possess good sense. For this reason as Cicero says in the book *On Duties*: Out of a desire to attain knowledge, renowned philosophers left their native country, going into exile never to return again afterwards.[45] And as Valerius reports in book eight, when Anaxagoras returned to his native country after a long period of study only to find the property he had left behind abandoned, he said, "I would not be saved if this had not been destroyed."[46] Socrates, too, said no one is rich enough to teach "unless he believes that he is poor and needy in learning."[47] For just as dust covers the eye and hinders sight, so also an abundance of goods extinguishes the light of reason*. And so the *Policraticus** says beautifully:

> For books, no hand stained with rust will be fit,
> Nor are moneyed minds able to have free time for books;
> The same person will not approve of both money and books.
> Money seekers cannot be identical with book lovers:
> No single house, believe me, contains both.[48]

41 Cf. Gen 19:11.

42 Cf. Prov 3:32.

43 The passage from "or his study" to "Gomorrah" is taken from John of Salisbury, *Policr.*, 7.13, ed. Webb, vol. 2, 147–48.

44 Isa 1:10.

45 Cf. Cicero, *De off.*, 1.20.69, ed. Winterbottom, 28.

46 Valerius Maximus, *Fact. et dict. mem.*, 8.7.Ext.6, ed. Briscoe, vol. 2, 526 (cf. Tubach, *Index exemplorum*, no. 3292).

47 John of Salisbury, *Policr.*, 7.13, ed. Webb, vol. 2, 151.

48 Ibid., Entheticus.269–71, 281–82, in *Policraticus I–IV*, ed. Keats-Rohan, CCCM 118:18 (cf. Walther, *Proverbia*, nos. 18929 [the first two verses] and 19135 [the last two verses]).

When Socrates was on the way to Athens in order to study there, he threw not a small weight of gold into the sea and said, "Oh evil riches full of worries; let me drown you, lest I be drowned by you!"[49] In book six of *On Architecture*, Vitruvius* relates that the Socratic philosopher Aristippus was shipwrecked and washed up on the Rhodians' shore. When he saw geometrical figures in the sand, he joyfully exclaimed to his companions, "We can hope for the best! I see the traces of humanity." And he immediately went up to the town of Rhodes, made his way to the school, disputed with the professors of philosophy there, and was provided with a great number of gifts so that he could not only provide himself with necessaries, but also give those who were with him clothing and other things essential for sustaining life. When his companions wanted to return to their native country and asked him what he wanted them to report at home, he said, "When you arrive in our native land, tell the people that it's necessary to furnish their children with those types of possessions and travel provisions that can survive if they're shipwrecked, and if those who possess them survive with this kind of assistance, it provides such protection for life that they can't be harmed by any change in affairs of state, or the destruction of war, or any other kind of adverse fortune."[50]

Behold, you have here a philosophical lesson and at the same time one that leads to salvation urging us to the acquisition of wisdom rather than to the acquisition of money, according to the passage in Wisdom 7: "I valued wisdom above empires and thrones, and I said riches were nothing in comparison with this, since all the gold in the world in comparison with this is only a little sand, and silver will be considered as dirt in view of this."[51] On this topic Theophrastus also says: "Among human beings, only the one who is learned will not be thought of as a stranger when he is in foreign lands. Even if he has lost his relatives, he will not be without indispensable friends, but rather in every state he will be counted as a citizen and a friend, and without fear he can disregard the hard calamities of fortune. But someone who is fortified not by knowledge, but by riches, treads on slippery paths, supporting himself with feeble and unstable assistance."[52] For even when Fortune bestows every one of her rewards, she easily takes them away. But the learning that has been joined to the soul remains with one permanently until the end of life. Hence, the foolishness of some moderns is clear who judge all knowledge a useless effort unless money results from it, although it is evident according to the judgment of the wise man that knowledge should be

49 Cf. Tubach, *Index exemplorum*, no. 3366.

50 Vitruvius, *De architectura libri decem*, 6.1–2, ed. F. Krohn (Leipzig, 1912), 120–21.

51 Wis 7:8–9.

52 Vitruvius, *De architectura*, 6.1–2, ed. Krohn, 121.

sought more than health and all beauty, Wisdom 7: "More than health and splendor,"[53] and so on.

On the chastity and abstinence of students*.

The third thing which hinders the required disposition of the eye is the fluid of carnal desire. For this reason Aristotle says in *On Animals*, 12, that birds having a pure and temperate fluid in their eyes fly very high and have acute sight,[54] as is clear from the eagle which flies higher than other birds due to the purity of its eyes and looks at the sun with an unblinking gaze.

In similar fashion, those possessing a purity of mind and flesh fly with wings of virtue on a loftier level and contemplate heavenly matters more clearly. For this reason, then, due to the sure token of his virgin purity, John the Evangelist flew higher than others like a heavenly eagle when he drank the rivers of wisdom from the well-source of the Lord's heart.[55] Paul, too, speaking of his own purity in 2 Timothy 1, says: "I give thanks to my God, whom I serve with a clear conscience,"[56] and as if he had already flown up to the heavenly spheres, he says: "Our dwelling-place is in the heavens," Philippians, chapter three.[57] Thus, he flew up on high – his sight was most acute – when he was caught up "to the third heaven," i.e., to the third order of vision, so that he could contemplate heavenly matters with his intellect, 2 Corinthians 12.[58] And Cassiodorus says that Paul arrived at such wisdom because of the purity of his flesh.[59] For since wisdom is "more beautiful than the sun," as is said in Wisdom 7,[60] it requires a clean lodging place. Gregory Nazianzen relates that when he was studying in Athens as a youth, two very beautiful maidens appeared to him when he was asleep, and they approached him and sank into his embrace. As a chaste man, however, he blushed with shame and, being not a little surprised, said, "But who are you?" And they said, "We're two maidens who're very close to you and whom you love. I'm Chastity," said the one. "And I'm Wisdom," added the other, "and we've come to you, most chaste youth, desiring to live with you, since you've prepared a holy

53 Wis 7:10. The full verse reads: "I loved knowledge more than health and splendor."

54 Cf. Avicenna, *De animalibus*, 12.11, in *Opera*, 49va; cf. Aristotle, *De part. animal.*, 2.13 (657b25–28), ed. Langkavel, 49.

55 Cf. John 7:38.

56 2Tim 1:3.

57 Phil 3:20.

58 2Cor 12:2.

59 Unidentified.

60 Wis 7:29.

dwelling-place for us in your heart and body."[61] For wisdom* imparts itself to holy souls and detests an unclean dwelling-place. In an image of this matter, one reads in Exodus 19 that as Moses was about to receive the law, he instructed the children of Israel to wash their clothing and abstain from women.[62]

Since the "law is immaculate,"[63] as the Psalm says, and "the words of the Lord are chaste words,"[64] divine wisdom requires an immaculate and chaste student. For this reason, Hugh says in the *Didascalicon*: "Unpraiseworthy is wisdom that is stained by a shameless life."[65] And in 1 Samuel 21, before Ahimelech* the priest gave the bread of the presence to David and his young men, he carefully asked, "Are the vessels of the young men clean, especially from women?"[66] In this passage we are given to understand mystically that they who desire to be fed with the bread of wisdom and understanding by Christ, the high priest, must be zealously chaste. And for this reason according to Augustine in book 12 of *On the Trinity*, Plato relates that "when a certain boy was asked about geometry, he replied as if he were the most skilled expert on the subject."[67] Augustine also says in book 8 of *The City of God* that Socrates was the first among the philosophers to direct philosophy toward moral phenomena, while all of his predecessors had been more interested in natural phenomena. Augustine says in the same passage that Socrates did this "so that the mind, freed from the oppressive weight of bodily desires," could ascend to a knowledge of divine matters, where it could investigate the causes of all things as if at their source.[68] For just as a dampened candle can barely be lit, so in a heart filled with carnal desire the light of instruction can barely

61 Cf. Aldhelm, *De virginitate*, 27, ed. Rudolf Ehwald, in *Opera*, MGH, Auctores antiquissimi, 15 (Berlin, 1919; reprint 1961), 262–63; cf. Rufinus of Aquileia, *Praefatio in Gregorii Nazianzeni orationes*, ad init., ed. August Engelbrecht, in *Opera*, CCSL 20 (Turnhout, 1961), 255 (cf. Tubach, *Index exemplorum*, no. 5367).

62 Cf. Exod 19:10–15.

63 Ps 18:8.

64 Ps 11:7.

65 Hugh of Saint-Victor, *Didascalicon*, 3.12, ed. Charles Henry Buttimer, The Catholic University of America Studies in Medieval and Renaissance Latin, 10 (Washington, DC, 1939), 61 (PL 176:773C).

66 Cf. 1Sam 21:1–6.

67 Augustine of Hippo, *De trin.*, 12.15.24, ed. Mountain and Glorie, CCSL 50:377–78. Augustine refers to Plato to illustrate the Platonic notion that souls lived before they entered their present bodies so that learning is actually remembering what the soul already knew. Peter uses the story to demonstrate that a child's innocence and chastity are rewarded with knowledge.

68 Augustine of Hippo, *De civ. Dei*, 8.3, ed. Dombart and Kalb, CCSL 48:218, 219.

be sustained, and if it happens to be received there, it is quickly extinguished by the wind of vanity. And just as the bodily eye cannot see clearly if it is bleary or covered with dirt, so also for the eye of the heart to contemplate the uncorrupted divine essence or knowledge, it must be purified of the sinful fluid of carnal desire. Therefore, it says in Second Kings 6: "The Lord opened the eyes of the young man, and he saw, and behold, the mountain was full of horses and chariots of fire in a circuit around Elisha."[69] For the perfect contemplation of wisdom cannot endure the foulness of unrestrained desire. For this reason Jerome says in *Letter 70 to the Monk Rusticus*: "Love the knowledge of the Scriptures and you will not love the vices of the flesh."[70]

Moreover, the fluid of carnal desire not only dulls and destroys the required disposition of the inner eye as far as lust is concerned, but also as far as gluttony is concerned. For as Seneca says: "A keen mind is dulled by great amounts of food."[71] For this reason the Psalm says: "They were confused and staggered about like a drunkard, and all of their wisdom,"[72] and so on. And Jerome, *Against Jovinian*: "We cannot pay attention to wisdom if we are thinking about a fully laden table."[73] For as the same author says in *Letter 56*: "It is not characteristic of one and the same person to have a taste for wine and to understand the prophets and apostles."[74] And Isidore, *On the Highest Good*: "Those who indulge in an excess of food, the more they feed their belly, the more they dull their mental understanding. For the Greeks have said: 'Subtle understanding cannot arise from a fat belly.'"[75] And therefore Solomon says in Ecclesiastes 2: "I considered how to withdraw my flesh from wine so that I could turn my heart to wisdom."[76] Therefore, it is said in Daniel 1 that to the young men who abstained from wine and did not want to fill themselves with foods fit for a king "God gave knowledge and understanding of every book and wisdom, but to Daniel he gave the comprehension of all visions and dreams."[77] And in Daniel 10 it is said: "In those days I, Daniel, mourned for three weeks, and I did not eat good-tasting bread, and meat and wine

69 2Kgs 6:17.

70 Jerome, *Ep.*, 125.11.2, ed. Hilberg, CSEL 56:130. See also below, chapt. 11.6, p. 141.

71 Seneca, *Ep. mor.*, 15.3, ed. Reynolds, vol. 1, 39.

72 Ps 106:27. The complete verse ends: "... and all of their wisdom had been devoured."

73 Jerome, *Adv. Iov.*, 2.11, PL 23:314B.

74 Jerome, *Ep.*, 61.3.2, ed. Hilberg, CSEL 54:579.

75 Isidore of Seville, *Sent.*, 2.42.6, ed. Cazier, CCSL 111:184 (PL 83:648A).

76 Qoh 2:3.

77 Dan 1:17; cf. Dan 1:4.

did not enter my mouth."[78] And afterwards [the archangel] Michael was sent to him and he revealed hidden things.[79]

Thus, since sobriety at the table and the study of wisdom are connected and have been joined together in such a strong alliance, it is not fitting for academics to make time for sumptuous dishes and to be eager to drain off goblets. For this reason a certain student wrote as follows to his companion who had abandoned his studies and was spending his time in nightly revels: "You who once neglected everything for educational duties and books have now transformed booklets into goblets and have turned inscribing into imbibing, and you are now proclaimed an infamous drinker, you who were formerly most renowned as a debater. You are more eager for goblets than for booklets, you find more leisure time for nightly revels than for reading, you are more eager for salmon than for Solomon. This is not 'a change made by the right hand of the Most High.'"[80] It is clear, then, how pertinent it is to the instruction of scholars that the required disposition of the visual organ is necessary for sight, and how this disposition is thought to be hindered by three things.

*It follows next that scholars should entrust their books not so much to a book bag as to their memory**.

Second, the presence of an object is required for sight. What a student hears from his teacher is not enough, but he must diligently pore over his books. For two senses are devoted to the acquisition of knowledge, namely sight and hearing. Therefore, it is said in Proverbs 10*: "The hearing ear and the seeing eye, the Lord has made both of them."[81] And therefore Augustine says in the first book of *Against the Academics* that without the work of scribes, the breeze* scatters students' efforts to understand like the winds blow sand.[82] Therefore, Jerome says in *Letter 41*: "If

78 Dan 10:2–3.

79 Cf. Dan 10:5–12, 13.

80 Cf. Peter of Blois, *Epistulae*, 7, PL 207:21B. The last sentence is taken from Ps 76:11. The passage depends on a series of near homonyms to emphasize the irony of the friend's transformation, e.g., he has transformed *codices* (booklets) into *calices* (goblets), he is more interested *in salmone* (salmon) than *in Salomone* (Solomon), etc.

81 Prov 20:12.

82 Cf. Augustine of Hippo, *Contra academicos libri tres*, 1.1.4, ed. William M. Green, in *Contra academicos; De beata vita; necnon De ordine libri*, CCSL 29 (Turnhout, 1970), 5. For the breeze scattering (*aure decerpunt*) one's efforts, see Virgil, *Aeneid*, 9.312–13, ed. and trans. Henry Rushton Fairclough, 2nd rev. ed., ed. George Patrick Goold, vol. 2, LCL 64 (Cambridge, MA, 1935; reprint 1974), 132–33 (*aurae / … discerpunt*).

only I could own books containing every treatise; that way my diligent study could compensate for my sluggish intellect!"[83] But some people want to own many beautiful and well-bound books without wanting to study them. And such people can be fittingly symbolized by Herod, whose name means "glorying in skins,"[84] for they glory in goat and calf skins. However, they should fear what is said in Job, chapter two: "Skin for skin,"[85] so that in fact for the goat skin which they "withhold in injustice" their own skin is sent to hell to be burnt. For as is said in Romans, chapter one, such people "withhold God's truth in injustice,"[86] since they ought to give the books they do nothing with to poor scholars who gladly would learn more. These people do not attain the fruit of knowledge since they rarely or never study the books which they keep firmly closed because of their beauty. And therefore it is not surprising if they understand very little in their books, for even though they own them, they keep the books under wraps from themselves.

For sight, as I said, the presence of an object is required, and therefore concerning these kinds of people the passage can be recited as one reads in Isaiah 29: "The vision of all this will come to you like the words of a sealed book. When people give it to someone who understands the letters," i.e., to someone who is thought to understand them, though in fact he understands nothing, "they will say, 'Read this book,' and he will respond," that is to say, he will have to respond, "'I cannot, for it is sealed.'"[87] About these kinds of students someone has said: "In trifling matters they are subtle, in important ones, dull and stupid; and lest they seem to have accomplished nothing when they return home, they put together large volumes made of calf skins, with wide margins, and they

83 Jerome, *Ep.*, 84.3.5, ed. Hilberg, CSEL 55:124.

84 In the Middle Ages, the name *Herodes* (in Latin) was commonly connected with animal skins, the material used to make the parchment and binding of books. See, for example, Anonymous, *In Matthaeum*, 2.1, ed. Bengt Löfstedt, CCCM 159 (Turnhout, 2003), 7: *Herodes interpretatur 'pellicius,'* an interpretation found in a number of other works, including Johannes Scot Eriugena's translation of Maximus Confessor and Thomas Aquinas' *Catena aurea in quatuor Evangelia, Expositio in Marcum*; cf. also Hermannus de Runa, *Sermones festivales*, 95.3, ed. Edmund Mikkers, Joseph Theuws, and Roland Demeulenaere, CCCM 64 (Turnhout, 1986), 445: *Herodes siquidem pellis gratia interpretatur*; Adam Scotus, *Sermones*, 34.12, PL 198:316C: *Herodes pellis gloria dicitur*.

85 Job 2:4.

86 Rom 1:18. It should be noted that Peter bequeathed his own appreciable collection of some 120 volumes to the library of the Sorbonne. Many of the volumes still survive, especially in the manuscript collection of the Bibliothèque nationale de Paris.

87 Isa 29:11.

bind them with beautiful, reddish skins, and thus they return to their native country with a wise book bag and a foolish mind. What sort of knowledge is that which a thief can steal,[88] a mouse chew on, a worm destroy, water damage, fire burn?"[89] When one of a number of such people was returning to his native land once with a large number of books, it happened that the pack animal[90] carrying his books fell into the water, and thus he lost everything. A certain scholar was walking behind him on foot, a pauper in terms of books, but rich in knowledge, who commended what he had heard at school not to his books but to his memory. And when he observed what had happened, he composed these verses and recited them to the student who had lost his books and his knowledge at the same time:

> To your heart, not a page, you should commit what you have skillfully learned:
> If the page perishes, wisdom still goes on with you.[91]

And for this reason Boethius says in the first book of *The Consolation*, prose 5: "It is not the walls of your library that I need, adorned with ivory and glass, but the interior of your mind, in which I have deposited not books, but that which makes books valuable – the meanings of books that used to be mine."[92] And Seneca says: "It does not matter how many books you have, but how well you have them stored up in your memory."[93]

That a separation from God through students' sinning destroys a knowledge of divine matters.*

Third, a proportionate amount of distance is required for sight. For if a visible object is placed directly on the eye, or if it is very distant from the eye, it will not be seen.[94]

88 Cf. Matt 6:19.

89 William of Conches, *Dragmaticon Philosophiae*, 3.1.2–3, ed. Italo Ronca, CCCM 152 (Turnhout, 1997), 55–56.

90 *Pr1* uses the unusual form *saccinarius* (laden with sacks) here, based on *saccus* (sack), and analogous with the adjective *sarcinarius* (relating to burdens). The latter word is used by Julius Caesar in reference to pack animals. In *P7P8*, the word that occurs here is *summarius* = *somarius* (pack animal).

91 *Sententia communis* (Walther, *Proverbia*, no. 3460); for the retort to the student by the scholar who was walking behind him on foot, see Tubach, *Index exemplorum*, nos. 3292 and 3745.

92 Boethius, *Phil. cons.*, 1.pr.5.6, ed. Bieler, CCSL 94:13.

93 Seneca, *Ep. mor.*, 45.1, ed. Reynolds, vol. 1, 115.

94 Cf. Roger Bacon, *Persp.*, 1.8.1, ed. and trans. Lindberg, 108–10. See above, chapt. 8.5, p. 92.

In similar fashion, if a scholar is very distant from anything to be learned because of hatred, or if he undertakes to learn it with an excessive amount of love, he does not see it nor does he judge it correctly. For love and hate pervert judgment. And Seneca says: Judgment perishes when the matter affects our desire.[95] Hence, it happens sometimes to many of us that our own deeds, though not very worthy, appear better than the deeds of others, although theirs are far superior to ours. For:

Everyone defends what he himself creates.[96]

As Petrus Alfonsi says: "By nature each person takes pride in his own song, whatever its quality, just as an ape also does in his offspring."[97] For an ape thinks that its own offspring surpasses all others in beauty, even though it holds last place in its ugliness. And for this reason the ape is depicted as having said about its son:

In my judgment he's better than everyone.[98]

Thus, the poet says as well:

An author praises his own work.[99]

And again:

Every scholar is passionate about his own studies.[100]

And just as a person judges his own deeds falsely from time to time since they are very near to him through love, so too will he judge others' deeds falsely if they are distanced from his affections* through hatred or ani-

95 Cf. Publilius Syrus, *Sent.*, N.11, available at http://www.thelatinlibrary.com/syrus.html (accessed December 10, 2010): "Nil rationis est, ubi res semel in affectum venit."

96 Ovid, *Ex Ponto*, 4.1.28, ed. Richmond, 84 (cf. Walther, *Proverbia*, no. 25801a).

97 Cf. Petrus Alfonsi, *Disciplina clericalis*, fabula 6, ed. Alfons Hilka and Werner Söderhjelm, Acta Societatis Scientiarum Fennicae, 38.4, Pars 1 (Helsingfors, 1911), 13. For the ape's pride in the beauty of its offspring, see Janson, *Apes and Ape Lore*, 41 and n54.

98 Avianus, *Fab.*, 14.14, ed. and trans. Duff and Duff, 704–05.

99 Ovid, *Ex Ponto*, 3.9.9, ed. Richmond, 80 (cf. Walther, *Proverbia*, no. 1677).

100 Ibid., 1.5.35, ed. Richmond, 16.

mosity. For as Gregory says: The intellect does not easily grasp what the affections abhor.[101]

The proportionate amount of distance required for sight can also be explained in another way: From the fact that a moderate amount of distance is required for vision (since a visible object is not seen when it is either too far away or too close) one can say that just as this is observed in corporeal vision, so too is a proportionate amount of distance required for spiritual vision. For a separation from God through faithlessness and committing many sins destroys a knowledge of divine matters, Isaiah 59: "Your iniquities have made a separation between you and your God, and your sins have hidden his face from you."[102] And Augustine says: "He errs who thinks he understands the truth, if he has been living immorally up to now."[103] This [spiritual sight] is likewise hindered when someone presumptuously thinks he is on intimate terms with what belongs to God and investigates God's sovereignty,[104] Proverbs 23: "Set limits to your knowledge, and do not raise your eyes to riches which you are unable to have, since they will make wings for themselves like an eagle and fly up to the heavens."[105] And in the Song of Songs, chapter 6: "Turn your eyes from me, for they have made me fly away."[106]

The fourth lesson. That one should be zealous about the kinds of knowledge that have substance, not those that are baseless, and so on.*

Fourth, a density in the visual object is required for vision. For this reason air, which is not a solid body, is not an object of sight.[107] By this fact scholars who are investigating what the intellect can visualize are taught that they should be zealous about the kinds of knowledge that have substance, not those that are baseless. For it is said in Wisdom, chapter thirteen: "All those are deluded in whom there is no knowledge of God."[108]

101 Unidentified.

102 Isa 59:2.

103 Augustine of Hippo, *De agone christiano*, 13.14, ed. Joseph Zycha, in *De fide et symbolo; De fide et operibus; De agone christiano; De continentia*, CSEL 41 (Vienna, 1900), 118.

104 Peter is indebted to Roger Bacon, *Persp.*, 3.3.2, ed. and trans. Lindberg, 326, for much of this paragraph.

105 Prov 23:4–5.

106 Cant 6:4.

107 Cf. Roger Bacon, *Persp.*, 1.9.1, ed. and trans. Lindberg, 124.

108 Wis 13:1.

Knowledge of God is the Holy Scriptures, through which someone knows God and also himself. And the blessed Bernard says in one of his sermons: "When a certain saint prayed, he said, 'God, let me know myself, let me know you.' A brief prayer, but pious."[109] For this is the true philosophy and, even beyond that, the perception absolutely necessary for salvation. Indeed, from the former of these, fear and humility are conceived; from the latter, hope and charity are generated. And Augustine says in *On the Trinity*: "Humankind commonly sets a high value on the knowledge of earthly and astronomical phenomena, but surely* those people are more correct who prefer knowing themselves to having that kind of knowledge. Indeed, the mind that knows its own weakness is more praiseworthy than one that has not investigated its weakness but searches out the paths of the stars just so it can know them, or one that retains the knowledge of the stars' pathways it has acquired already, though ignorant of the path it must take for its own salvation."[110] And Seneca says in *On the Liberal Arts*: It is silly to be zealous about measuring the dimensions of land masses and not to take the measure of oneself. And it is foolish to have free time for the harmony of musicians and not to be in harmony with one's neighbor or even with oneself.[111] And Bernard, in *Sermon 36 on the Song of Songs*: "The usefulness of knowledge depends on the manner in which it is known, so you should know in what order, with what zeal, and to what end you ought to know something*: in what order, so that you know first whatever is best for salvation; with what zeal, so that you know more passionately whatever pertains more ardently to love; to what end, so that you know not for the purpose of vainglory or empty curiosity or anything similar, but for the edification of yourself or your fellow human being."[112]

The Holy Scriptures are, as it were, the pharmacy of the Holy Spirit, in which one finds a wholesome ointment for every wound, an efficacious antidote for disease, a salubrious remedy for pain. And as Chrysostom says: "In the Holy Scriptures the ignorant person discovers what he ought to learn; the insolent one, what he ought to fear; the one laboring,

109 Bernard of Clairvaux, *Sermones de diversis*, 2.1, ed. Jean Leclercq and Henri M. Rochais, in *Opera* (Rome, 1970), vol. 6/1, 80 (cf. Augustine of Hippo, *Soliloquiorum libri duo*, 2.1.1, ed. Wolfgang Hörmann, in *Soliloquiorum libri duo, De inmortalitate animae, De quantitate animae*, CSEL 89 [Vienna, 1986], 45).

110 Augustine of Hippo, *De trin.*, 4.proem.1, ed. Mountain and Glorie, CCSL 50:159.

111 Cf. Seneca, *Ep. mor.*, 88.9, ed. Reynolds, vol. 1, 314.

112 Bernard of Clairvaux, *Serm. sup. Cant.*, 36.2.3, ed. Leclercq, Talbot, and Rochais, in *Opera*, vol. 2, 5.

his reward; the faint-hearted, his comfort."[113] Behold what great solidity there is in Holy Scriptures and how distant vanity is from them! And thus in Proverbs 22 the wise man, inviting us to the study of wisdom, says: "Behold, I have described it to you in three ways in thoughts and knowledge to show you the durability and eloquence of truth."[114] And therefore it is to this we ought to raise up the eye of our reason, as it were to a firm and solid object. And thus it is said in Second Peter, chapter one: "We have the prophetic word which endures longer and to which you would do well to pay attention."[115] Thus, according to Jerome's advice in *Letter 35 to Paulinus*: "Let us learn those things on earth, the knowledge of which may remain with us in heaven."[116] But Holy Scriptures, which are very substantial, are considered vain and empty by many who say the verses in Numbers 11: "Our eyes see nothing else but this manna"[117] (which is interpreted to mean "What is this?"[118]) and Numbers 21: "Our soul is nauseated by this food which is not filling."[119] But such people's vision is deceived, because holy doctrine contains the most substantial truth. It is said in Second Peter, chapter one: "For no prophecy was ever announced from the human will, but God's holy people have spoken, moved by the Holy Spirit."[120] And Proverbs 12: "Lips speaking the truth will endure forever."[121]

In secular disciplines, however, there is the vanity of multiple error. Therefore, it is said in Jeremiah 10: "The Lord says this, 'Do not learn according to the ways of the nations, for the laws of the peoples are vain.'"[122] And Titus 3: "But avoid foolish and inane questions, for they are useless and vain."[123] It is related in Genesis 11 how communication in a unified language was taken away from those who were presumptuous when the tower of Babel was being erected presumptuously into the heavens, and how a partition of peoples followed a separation of lan-

113 Ps.-John Chrysostom, *Opus imperfectum in Matthaeum*, Hom. 41, PG 56:862. See also Thomas Aquinas, *Catena aurea, Expositio in Matthaeum*, 22.1, in *Catena aurea in quatuor Evangelia*, ed. Angelico Guarienti (Turin, 1953), vol. 1, 318.

114 Prov 22:20.

115 2Pet 1:19.

116 Jerome, *Ep.*, 53.10.2, ed. Hilberg, CSEL 54:464.

117 Num 11:6.

118 Jerome, *Lib. interp. Hebr. nom.*, Apoc., ed. de Lagarde, CCSL 72:160.

119 Num 21:5.

120 2Pet 1:21.

121 Prov 12:19.

122 Jer 10:2–3.

123 Tit 3:9.

guages.[124] So also while the philosophers raised to the heights the devices of their intellect, prying into heavenly and divine matters with an illicit curiosity and presuming to do so by their own powers, the unity of truth, which is real and immutable and unfailing*, was removed from them. Enveloped by clouds of ignorance and given over to a reprehensible way of thinking, they were dispersed into sects of various types of errors, and "claiming that they were wise, they were made fools," Romans 1.[125] And therefore in their sayings there is sometimes too little which is substantial and too much which is vain, Ecclesiastes, chapter six: "There are very many words, and in discussions they possess a great deal of vanity."[126] And Jeremiah, chapter two: "They went after vanity, and they were made vain."[127] And one can say the verse in Isaiah 56 about them: "All his investigators are blind, seeing vain things."[128] And when they finally perceive this, they say, or they are able to say, the verse by the Psalmist: "Our eyes have grown dim with the lack of food,"[129] because in such areas of study one finds no spiritual nourishment. For this reason in *Letter 14 to Damasus*, speaking about the pods with which the prodigal son was not able to be sated,[130] Jerome says as follows: "And since one meets with no surfeit of truth in them, no nourishment of justice," those eager for the pods are left hungry for the truth and destitute of virtue.[131] And therefore it says in the Psalm: "Blessed is the man who makes the name of God his hope and who does not look after vanities and deceitful follies."[132] For these are the areas of study that make people insane, and for this reason it was said to the apostle Paul in Acts 26: "You are insane, Paul; your great book learning is making you insane."[133] Hence, no one should be too intent on secular studies. And so in *Letter 82 to Eustochium* Jerome relates about himself that because he was very devoted to books by Cicero, he was led before the tribunal of the eternal judge and punished there most severely.[134] It is related in the *Life of Hugh of Cluny** that when he was sleeping, he had a dream in which he saw a great number of ser-

124 Cf. Gen 11:1–9.
125 Rom 1:22.
126 Qoh 6:11.
127 Jer 2:5.
128 Isa 56:10.
129 Ps 87:10.
130 Cf. Luke 15:16.
131 Jerome, *Ep.*, 21.13.4, ed. Hilberg, CSEL 54:122.
132 Ps 39:5.
133 Acts 26:24.
134 Cf. Jerome, *Ep.*, 22.30.1–6, ed. Hilberg, CSEL 54:189–91 (cf. Tubach, *Index exemplorum*, no. 2774).

pents and wild beasts lying under his head. Waking up and investigating the matter carefully, he discovered under his pillow a book of tales that was filled with obscenities*, the rituals of the heathens. But after he threw the book out, he slept peacefully.[135]

And therefore a scholar ought to pray: "Turn away my eyes, lest they see vanity," that is to say, the vanity of secular studies; "in your way give me life."[136] This way is the Holy Scriptures or divine law, about which the Psalm says: "I have run in the way of your commandments."[137] This is the way that leads to heaven, Wisdom 6: "The desire for wisdom leads to the eternal kingdom."[138] Thus, each and every scholar should be devoted to this divine wisdom through his study and should dedicate to its service whatever he has learned from secular studies. For this reason, explaining what is found in Deuteronomy 21, Jerome says in *Letter 13 to Damasus*: "An image of secular wisdom is depicted in Deuteronomy in the figure of the woman about whom the divine voice commanded that if an Israelite wanted to marry her, he had to make her bald, cut her nails, and shave her hair. After she had been cleansed, she was able to be embraced by her spouse.[139] If we understand these words according to their literal sense, are they not ridiculous? But we, too, are accustomed to doing this when we read the philosophers, when books of worldly wisdom come into our hands: Whatever we find useful in them, we apply to our doctrine, but if we find anything useless dealing with idols, or romance, or worldly cares, we lop that off, we decree baldness for them, we cut it off like fingernails with the sharpest blade."[140] And therefore it should not seem senseless or inappropriate to anyone if words of philosophy or political science are granted entry into our teachings, especially when reason has required this. Nor should anyone try slanderously to find out from which academic faculty these words are taken as long as they are helpful for salvation, just as no one rightly asks which land produced an herb or which gardener took care of the land, as long as the herb is good for one's health. And therefore sometimes things can be taken from

135 *Epitome vitae sancti Hugonis abbatis Cluniacensis ... per Anonymum excerpta*, 6, in *Acta sanctorum Aprilis collecta, digesta, illustrata*, ed. Godefroid Henschen and Daniel van Papenbroeck (Antwerp, 1675), vol. 3 (29 April), 656D. The text of Hugh's life attributes the authorship of the book of pagan obscenities to "Maro," i.e., Virgil. Hugh (1024–1109) was installed as the abbot of the monastery of Cluny in 1049.

136 Ps 118:37.

137 Ps 118:32.

138 Wis 6:21.

139 Cf. Deut 21:10–13.

140 Jerome, *Ep.*, 21.13.5–6, ed. Hilberg, CSEL 54:122–23.

pagan books for the word of God because of their general educational value, for when he wanted to refute certain men, even the apostle Paul used the verses of the poet Epimenides*, Titus, chapter one.[141]

The fifth lesson, on diligent attention to and retention of what is taught, and the indefatigable love for the devotion to learning.*

Fifth, the application of the visual faculty is required for vision. For although the *species* of many visible objects have an effect on the mediating air and are arrayed about the eye and the visual spirit,[142] nevertheless of all those visible objects located about the eye, our vision often perceives only one, namely that one on which the soul has fixed its attention.[143] In the same way, when a hunting dog is firmly locked onto the scent of a stag by the application of its hunting instincts, it pursues that stag in such a way that it seems neither to see nor to perceive other stags that it happens to come across or that appear in its path.[144] Concerning this attention in seeing, it is said in Isaiah 48: "If only you had paid attention to my commandments!"[145] And the Lord urges us to do so in Deuteronomy 11: "Place these words of mine in your heart and set them between your eyes."[146] But there are many students who are like little monkeys: They always stay close to the instructor who is lecturing and they never leave the schoolrooms, and yet they do not pay any attention to the instructor's teachings. And so, although they remain in his classes for a long time, they know nothing. For study benefits a student only as long as he diligently pays attention to what he reads or hears and when he reflects on what he has heard or read and examines it internally in the cabinet or locker of his heart. To read otherwise and not to understand is

141 Cf. Tit 1:12, quoting a line from the *Cretica* by Epimenides of Knossos.

142 On the visual spirit(s), see above, chapt. 7.5, p. 54, n50.

143 The importance of intention in completing sight is not stated explicitly by either Roger Bacon or Alhacen. Bacon, however, maintains that "vision must occur through its species extended to the visible object" from the eye, which may be taken to imply that the intention of the viewer is an element in the completion of vision. See Roger Bacon, *Persp.*, 1.9.1, ed. and trans. Lindberg, 130. The same point is made by John Pecham, *Persp. comm.*, 1.46, ed. and trans. Lindberg, 128–30, who maintains that "rays [emitted by the eye] do have some effect on sight" (p. 129). See also above, chapt. 4, p. 14, where the processes of both intromission and extramission, which involves the cooperation of one's own *species* and visual power, are said to be necessary for vision to occur.

144 Cf. Tubach, *Index exemplorum*, no. 1708.

145 Isa 48:18.

146 Deut 11:18.

to be neglectful. Therefore, about the daily reading Bernard says to the brothers of Mont Dieu: "A part of it should be preserved in the belly of your memory which you will very faithfully digest, and when you recall it anew, you can ruminate on it again and again."[147] And Seneca says: "After you have gone over many things during the day, select one of them that you will chew over carefully that day."[148] And again: For those who go over their writings from every angle and have a purpose for each and every word, it is unavoidable that they will imprint things on their memory as often as they recall them to mind. For whoever wields a swift pen also has a fleeting memory.[149]

But there are many lazy students who do not want to apply their minds to studying, so they find a way to avoid the many tasks involved in studying. Indeed, as Seneca says: "It is stupid to study lazily."[150] And as long as a slothful cleric embraces a life without labor, he takes comfort in his ignorance. For when he hears that others are eagerly studying, he says, "Why should I lower myself to the labor of studying? Why should I be zealous for wisdom? For not all of those who are without wisdom wind up perishing." And then he says, "There are enough wise men in the world. There's no need for me to wear myself out and torment myself with learning. I've left early childhood behind and I'm approaching old age, so it wouldn't be possible for me to attain the great fruits of knowledge any more if I were to begin working on them now." In this way the slothful person thinks things over within himself and, at one and the same time, persists in laziness and ignorance. Nor does he perceive that the ancient enemy is suggesting these things for his destruction so that he will spend his entire life with nothing that is useful, but always live in sluggishness and neglect and likewise ignorance, and in the end he *will* perish. For he says, "The educated person dies together with the uneducated one; in the end I'll die repentant and I'll be saved as surely as someone with knowledge." Concerning such people a poet says:

> There are some who want to know and give no effort for it,
> Nor do they endure study and toil. This is the way of the cat:

147 William of Saint-Thierry, *Epistola ad fratres de Monte Dei*, 122, ed. Volker Honemann, in *Die "Epistola ad fratres de Monte Dei" des Wilhelm von Saint-Thierry. Lateinische Überlieferung und mittelalterliche Übersetzungen*, Münchener Texte und Untersuchungen, 61 (Munich and Zurich, 1978), 336 (cf. PL 184:327D).

148 Seneca, *Ep. mor.*, 2.4, ed. Reynolds, vol. 1, 3.

149 Cf. Seneca the Elder, *Controversiae*, 1.pref.18, ed. and trans. Michael Winterbottom, in *Declamations*, vol. 1, LCL 463 (Cambridge, MA, 1974), 16–19.

150 Seneca, *Ep. mor.*, 71.23, ed. Reynolds, vol. 1, 214.

> It wants the fish, but not the fishing. I do not speak to them,
> But only to those who enjoy the toil of learning as well as
> knowing.[151]

I had a companion who was able to stay up the entire night talking about idle and frivolous matters, but who nevertheless, as he frequently confessed to me, so detested his studies that if he were perhaps unable to sleep at some point, he would open a book and after reading four or five lines would fall asleep immediately. This man was clearly the opposite of the one about whom Cassian speaks in book 5 of the *Institutes* for monks. "When," he says, "we had traveled to the province of Egypt, we saw an elder named Machetes* who by his continual prayers had received from God this grace, that whatever was said to him during days and nights in spiritual conversation, he never fell asleep. However, if someone attempted to interject a disparaging or idle word, he immediately sank into sleep. And thus the pollution caused by the poison of speech could not even reach his ear."[152]

One reads that the philosophers of a former age were also not so lazy in studying, but rather were more like indefatigable warriors for wisdom who were remarkably attentive around the school of philosophy. Valerius Maximus relates in book 8 of the *Memorable Sayings and Deeds* that after the capture of Syracuse, Marcellus realized that his victory had been delayed for a long time chiefly by the inventions designed by the philosopher Archimedes. Nevertheless, he took delight in the philosopher's extraordinary skill and decreed that his life should be spared, reckoning there to be almost as much glory in an Archimedes kept unharmed as in a Syracuse overthrown. However, when a soldier broke into Archimedes' house to plunder it, then drew his sword, held it over Archimedes' head, and asked him who he was, this philosopher was sketching diagrams and his thoughts and eyes were intent on the floor. Because of his great desire to continue researching what he wanted to know, Archimedes did not want to disclose his name to the soldier, but rather said, with his hands held over the sand, "I beg of you; don't disturb this!" The soldier cut off his head, as if he were someone who disrespected the order of the victor, and Archimedes' blood ruined the designs of his diagram. And so it hap-

151 Geoffrey of Vinsauf, *Poetria nova*, 2027–30, ed. Edmond Faral in *Les arts poétiques du XIIe et du XIIIe siècle. Recherches et documents sur la technique littéraire du moyen âge* (Paris, 1962), 259 (cf. Walther, *Proverbia*, no. 30817).

152 John Cassian, *De institutis coenobiorum et de octo principalium vitiorum remediis libri xii*, 5.29, ed. Michael Petschenig, CSEL 17 (Vienna, 1888), 103–04 (cf. *Vit. pat.*, 4.34, PL 73:837B; cf. Tubach, *Index exemplorum*, no. 3911). Machetes is known only from his mention by Cassian.

pened that because of one and the same study, life was first granted and then taken away.[153] Thus, the same activity gave him life and took it from him. Jerome also relates in *Against Jovinian* that when the philosopher Antisthenes would not take on any students and was unable to force the persistent Diogenes to leave him, he finally threatened Diogenes with a cudgel so that he would go away. Diogenes is said to have bared his head and responded, "No club could be hard enough to keep me from loyalty to you."[154] How eagerly he chose to hear his teacher's words with his attentive ears if a cudgel could not drive him away! Symmachus also relates in his letters that unhappy news concerning Anaxagoras' son did not make this philosopher leave the debates.[155] Valerius also relates that the philosopher Solon* always burned with the greatest fervor for wisdom. On the last day of his life, when he was at the very point of dying and other philosophers – his friends – were sitting around discussing a particular question among themselves, he raised up his head, now thoroughly weighed down. And when he was asked why he did this, he responded, "So I can leave the world a happier man when I've learned what you're debating."[156] Thus, let lazy students blush with shame when they hear that even a dying man did not stop paying attention to his studies! One also reads in the chronicles that when the distinguished poet Virgil was asked how he composed his verses, he responded, "You ask about my method of composition? I give birth to my verses in the manner of a bear," he said, "for just as a mother bear delivers an offspring that is shapeless and without form and afterwards shapes and forms it by licking it, so the verses that I give birth to have a rough and imperfect appearance when they first come into being, but afterwards by working with them and rewriting them with the greatest care, I make them well-formed."[157] And therefore as a certain philosopher said: "Determination

153 Cf. Valerius Maximus, *Fact. et dict. mem.*, 8.7.Ext.7, ed. Briscoe, vol. 2, 526–27.

154 Jerome, *Adv. Iov.*, 2.14, PL 23:318B–C.

155 Cf. Quintus Aurelius Symmachus, *Epistolae*, 3.6.3, ed. Otto Seeck, in *Opera quae supersunt*, MGH, Auctores antiquissimi, 6 (Berlin, 1883), 72 (cf. Tubach, *Index exemplorum*, no. 202).

156 John of Salisbury, *Policr.*, 7.13, ed. Webb, vol. 2, 149; Valerius Maximus, *Fact. et dict. mem.*, 8.7.Ext.14, ed. Briscoe, vol. 2, 529–30 (cf. Tubach, *Index exemplorum*, no. 2968).

157 Cf. Vincent of Beauvais, *Spec. hist.*, 6.62, in *Speculum quadruplex*, vol. 4, 193b (cf. Suetonius/Donatus, *Vita Donati*, 22, ed. Colin Hardie, in *Vitae Vergilianae antiquae*, 2nd ed. [Oxford, 1966], 11). On Virgil's response, see Jacques Berlioz, "Virgile dans la littérature des *exempla* (XIIIe–XVe siècles)," in *Lectures médiévales de Virgile. Actes du Colloque organisé par l'École française de Rome (Rome, 25–28 octobre 1982)*, Collection de l'École française de Rome, 80 (Paris and Turin, 1985), 82–83, 106.

is the mother of the arts; neglect, the wicked stepmother of learnedness."[158]

Someone describing the life of a serious student said that he drinks from a spring, weeps during prayer, and sleeps on his books.[159] But today one finds many lazy students and few industrious ones. A certain layman who was a close acquaintance of mine once asked me in which faculty more students were enrolled at Paris, and I responded that this was in the Arts or in Logic. And he said, "On the contrary, rather in Music (*musica*), since more of them grumble about things (*mussitant*) than study." Speaking in an elegant meter, Architrenius describes the diligence and attentiveness of students in their study as follows:

> Exhausted, with both eye and mind sleepless, he drags out
> The nights in vigils with each of them, and ever-watchful he also inflames his eyes
> With the lamp and tortures them into bleariness.
> Thus, he presses into his books with the lamp of his eyes and mind,
> And resting his elbow on the book and his right hand* on his ear
> He contemplates what modern wisdom and that of the ancients has produced.
> Drinking it in now with his eyes, now with his mind, now gulping it down with both,
> With the former more, with the latter less; now in the furnace that is his intelligence
> He boils down the things he has read, and in his memory he binds them in his breast,
> Prescient, with a knot. Now with deep concentration he gnaws on whatever very knotty passage obstructs and
> Holds back his comprehension and keeps it from taking wing
> To the heights, but with the full exertion of his mind
> Grasping at what lies beyond it and the rewards of study, his eye burns
> And his head sinks to his chest, and he considers at length how to make the steep

158 Magnus Felix Ennodius, *Opera*, 124(*Dictio* 11).3, ed. Friedrich Vogel, MGH, Auctores antiquissimi, 7 (Berlin, 1885), 132 (cf. Gratian, *Decretum Magistri Gratiani*, 1.83.1.6, ed. Emil Ludwig Richter, 2nd ed., ed. Emil Friedberg, in *Corpus iuris canonici* [Leipzig, 1879], vol. 1, 294).

159 Cf. Ambrose of Milan, *De virginibus*, 3.4.15, ed. and trans. Franco Gori, in *Verginità e vedovanza, I*, Sancti Ambrosii Episcopi Mediolanensis Opera, 14/1. Opere Morali 2/1 (Milan and Rome, 1989), 220.

Paths accessible, and how to break through resisting doors
That deny him entrance, and very frequently he brings his
Eye, the tip of his finger, and the keenness of his mind back to the book.[160]

After this, he adds verses of the following kind "concerning the sleep of students fatigued by study":

No restfulness, this sleep! For the care needing to be calmed
Finds no quiet in repose, and with no peaceful slumber
The restless toil of study is without peace, and care remains awake even
In sleep: It assigns books and tasks to work on, the repose of
The student's mind is agitated, nor does the overwhelming worry
Yield to slumber*, but that uneasiness of mind returns
Which had kept the mind awake earlier. What the very sluggish intellect
Of the student had given up hoping for earlier when awake, deep night reveals with sleep*
As his teacher: Although on the outside the baggage of the prostrate body
Sleeps, within the mind its perceptive powers also stay awake.[161]

Finally, in these verses Architrenius describes the attentiveness of students in the presence of their professor:

He stirs up the flames of his intelligence, gives focus to his mental
Efforts, sharpens his mind, while his body attends to what his
Eyes transmit to his straining intellect, and he drinks in what his master says
With fertile ear and mind, and with both of them he is prepared:
He quickly gathers up the words that fall, and his field of vision and thought
Is focused on his teacher, fixed and watchful, and his ear, like a bridesmaid,

160 Johannes de Hauvilla, *Arch.*, 3.6.99–7.104, 3.7.107–10, 3.7.115–22, ed. Schmidt, 164–65; trans. Wetherbee, 67. The passage is taken from chapters devoted, respectively, to the bed in which the scholar sleeps and his study at night.

161 Ibid., 3.9.194–99 (and the chapter title), 3.10.201–02, 3.10.204–07, ed. Schmidt, 167–68; trans. Wetherbee, 71. The verses are taken from chapters devoted, respectively, to the sleep of a scholar who has been exhausted from studying and the way in which he is vexed by his dreams.

Weds his beloved and desired Minerva to his mind.[162]

Note that there are seven keys to being a philosopher, which the *Policraticus* sets out in these verses:

> A humble mind, a passion for investigating things, a settled life,
> A quiet study, poverty, dwelling off the beaten track.[163]

And the seventh key is affection for one's teachers. For this reason Quintilian also says:

> Teachers are accustomed to making those difficult passages that must be read accessible to many*.[164]

The sixth lesson. That one must zealously dwell on reading, nor should one leave off reading before death.

Sixth, a duration of time is required for vision, for as Ptolemy proves in the second* book of *On Appearances*, when a visible object is presented to the sight in a split second, it is not perceived distinctly unless it is first given some amount of consideration, which requires a duration of time that can be measured.[165] In this fact the student is taught that he should dwell on

162 Ibid., 3.15.291–97, ed. Schmidt, 171; trans. Wetherbee, 77. The passage is taken from Johannes's description of a student's behavior in the presence of his teacher. It is fitting that Minerva is wedded to the student's mind since she is the goddess of wisdom in classical mythology.

163 John of Salisbury, *Policr.*, 7.13, ed. Webb, vol. 2, 145.

164 Ibid.; the third verse follows on immediately after the preceding two in the *Policraticus*, with no intervening sentence and no mention of Quintilian. The enumeration of the keys of being a philosopher is attributed to *senex Carnotensis* in the *Policraticus*. The three verses, again without mention of Quintilian, are also found in Hugh of Saint-Victor, *Didascalicon*, 3.12, ed. Buttimer, 61 (PL 176:773B).

165 Cf. Roger Bacon, *Persp.*, 1.9.2, ed. and trans. Lindberg, 132, who also refers to book 2 of Ptolemy's *Optics*. See Ptolemy, *Optica*, 2.130, ed. and trans. Lejeune, 77–78; *Ptolemy's Theory of Visual Perception*, trans. Smith, 123. Cf. also John Pecham, *Persp. comm.*, 1.53, ed. and trans. Lindberg, 134 (see Schleusener-Eichholtz, "Naturwissenschaft," 295n177); Alhacen, *De aspect.*, 3.3 [3.11], ed. and trans. Smith, in *Alhacen's Theory of Visual Perception*, vol. 1, 289 (Latin text), vol. 2, 591 (English translation). The emphasis in this passage of Ptolemy's work is on the illusion of speed when, for example, an object passes over the entire field of limited vision (a narrow aperture, for instance), or vanishes before it reaches the

his reading and not rush through it quickly, so that he can achieve a knowledge of truth. For this reason Bernard says to the brothers of Mont Dieu: "Study differs from reading as friendship differs from hospitality or a deep affection from a chance greeting."[166] And Gregory in book 5 of the *Morals*: "When someone seeks an understanding of the truth, he penetrates it only by hard work."[167] One reads in Genesis 32 that the patriarch Jacob engaged in a wrestling contest and he sprained the tendon in his thigh. But Jacob did not quit the wrestling match until he obtained the angel's blessing.[168] Using his example, we too, on the road of our pilgrimage, ought to grasp Scriptures as if they were an angelic messenger of the divine will and wrestle with them until sunrise, i.e., until an understanding of truth begins to dawn in our mind and shine light on it, Ecclesiasticus 51: "My mind has wrestled with wisdom."[169] In this contest the tendon in the thigh is sprained, since the desire of the flesh is weakened in human beings by the study of wisdom, Jerome: "Love the knowledge that results from scholarship and you will not love the sins of the flesh."[170] From the fact that a student wrestles with wisdom using his intellect and his will as if they were his two arms, he obtains high praise as a blessing, Proverbs 4: "Get wisdom, get prudence; seize it and it will exalt you, you will be glorified by it when you embrace it."[171] For as Gregory says in the *Morals*: "Sometimes a studious person is punished by being slow to understand so that the more he exerts himself in his efforts to learn, the greater will be the rewards of the compensation he earns."[172] And concerning this wrestling match for wisdom a poet says:

> Sometimes a studious mind, anxious, fights like a boxer:
> He should stand his ground to the end; the passage of time may
> yield what the present does not.[173]

---

edge of the visual field. Alhacen, too, stresses that when an object moves quickly, and in a limited field of vision, "it will not be correctly perceived by sight." Bacon introduces into this discussion the necessity of a duration of time to allow for visual judgment to correct what is not seen distinctly because it moves quickly, and this is the point that Peter emphasizes, as well, and takes as the basis of his moralization, when something appears to the eye all of a sudden.

166 William of Saint-Thierry, *Ep. ad fratres de Monte Dei*, 121, ed. Honemann, 336 (cf. PL 184:327D).

167 Gregory the Great, *Mor. in Iob*, 4.praef.1, ed. Adriaen, CCSL 143:158.

168 Cf. Gen 32:22–32.

169 Sir 51:25.

170 Jerome, *Ep.*, 125.11.2, ed. Hilberg, CSEL 56:130. And see above, chapt. 11.1, p. 124.

171 Prov 4:5; Prov 4:8.

172 Gregory the Great, *Mor. in Iob*, 6.11.13, ed. Adriaen, CCSL 143:293.

173 *Sententia communis* (Walther, *Proverbia*, no. 14710).

By the fact that a duration of time is required for vision we are taught to show perseverance in reading. We are also instructed by this fact to embrace the study of wisdom at all times. For in *Letter 25* Jerome says: "It seems to me there is no age at which it can be too late to learn, since even if it is more fitting for an elderly person to teach than to learn, nevertheless it is better for him to learn than to teach in a state of ignorance."[174] And Augustine to Volusian: "Christian writings are so profound that I would always be learning something new from them even if I had undertaken to learn more about them from the beginning of childhood until a ripe old age with the maximum amount of free time, the greatest zeal, and with a more astute mind."[175] And therefore the wise man, speaking obviously about wisdom, says in Wisdom 8: "I loved her and sought her out from the time of my youth and sought to take her for my wife."[176] For a man should take wisdom like a wife, not like a mistress, so that he does not stay with her for a while and afterwards leaves her behind, but is joined to her in an unbreakable bond. One reads that the philosophers formerly acted in this way, for indeed they are worthy of being remembered for many reasons. Whence, Valerius Maximus says in book eight: "It is reported that when Plato lay dying at the age of 81, he kept Sophron's books under his head. Thus, not even at his final hour was he unoccupied with acquiring knowledge through study, but rather the conclusion of applying himself to philosophy came only with the conclusion of his life. In this fact he shows that some people do not have enough time, while for others there is abundant time."[177] And in the *Institutes of Oratory* Quintilian says: "The love of knowledge and the practice of reading are not confined to one's student days, but continue for the length of one's life."[178]

174 Augustine of Hippo, *Ep.*, 166.1.1, ed. Goldbacher, CSEL 44:546; also edited in Jerome, *Ep.*, 131.1.1, ed. Hilberg, CSEL 56:202.

175 Augustine of Hippo, *Ep.*, 137.1.3, ed. Goldbacher, CSEL 44:99–100.

176 Wis 8:2.

177 Valerius Maximus, *Fact. et dict. mem.*, 8.7.2, 8.7.Ext.3, 8.7.Ext.5, ed. Briscoe, vol. 2, 519, 524, 525. Valerius asserts that the conclusion of philosophy comes at the end of life in a passage devoted not to Plato, but to Carneades of Cyrene, a skeptic philosopher associated with the Platonic Academy; the last sentence of the series of quotations from Valerius comes from a passage in which Valerius praises Cato the Younger, the Roman statesman, not Plato.

178 Quintilian, *Institutionis oratoriae libri xii*, 1.8.12, ed. Ludwig Radermacher, rev. ed., ed. Vinzenz Buchheit (Leipzig, 1965), vol. 1, 54.

The seventh instruction. That students lack the light of grace, which is God. And on the threefold eye, and so on.*

Seventh and last, the diffusion of light illuminating the medium is required, for nothing is seen in the dark. For this reason it is proven in *Perspectivist Science* that nothing is seen except by a light shining on the visible object that is also multiplied from this object onto the eye.[179]

Thus, in order for our spiritual eye to perceive some truth, it is necessary that the brilliance of the sun of justice be shining on it. For God is "the true light which illuminates every human being coming into this world," John 1.[180] And just as without a corporeal light there is no corporeal vision, so without this other light there is no spiritual vision. For it is God "who teaches all knowledge," as it is said in the Psalm.[181] And Augustine: "Whoever teaches people's hearts possesses his chair in heaven."[182] For only God teaches a human being as an efficient cause, but one human being teaches another in a hortatory way. Augustine proves this in *The Teacher* by this triple sign: First, since a person can achieve knowledge without a human being as his instructor, it is therefore necessary that he have someone else as an instructor, and that is God, who instructs internally, according to the Psalm: "Let me hear what the Lord God will speak within me."[183] Second, since sometimes by teaching things that he does not understand, one person teaches another, which could not happen if the presence of a learned person were required to bring about knowledge, since no one can give what he does not possess. Third, since after an instructor has accomplished whatever he can, a student does not emerge in full possession of knowledge immediately, but rather knowledge is produced in him only after he thinks things over

179 Cf. Roger Bacon, *Persp.*, 1.8.1, ed. and trans. Lindberg, 108, who also refers to Alhacen here, and Alhacen, *De aspect.*, 1.5 [4.27], ed. and trans. Smith, in *Alhacen's Theory of Visual Perception*, vol. 1, 10 (Latin text), vol. 2, 347 (English translation). For the explanation of *multiplicatio*, by which light generates light through the transparent medium from the visible object to the eye, see Roger Bacon, *De multiplicatione specierum*, 2.1, ed. and trans. David C. Lindberg, in *Roger Bacon's Philosophy of Nature. A Critical Edition, with English Translation, Introduction, and Notes, of De multiplicatione specierum and De speculis comburentibus* (Oxford, 1983), 90–95.

180 John 1:9.

181 Ps 93:10.

182 Augustine of Hippo, *In Epistolam Joannis ad Parthos tractatus x*, 3.13, PL 35:2004.

183 Ps 84:9.

and, as it were, consults his internal light of reason.[184] Therefore, in *Letter 34 to Paulinus* Jerome says: "If all things that have been written are not made accessible by him 'who has the key of David – which opens, and no one closes; closes, and no one opens'[185] – they will not be unlocked and revealed by anyone else."[186] And Bernard in one of his letters: "You deceive yourself, my son, you deceive yourself, if you think you can find among secular masters a knowledge that only the disciples of Christ obtain as a gift from God. For this is something that lecturing does not teach, but anointing."[187] This is why the Apostle says: "His anointing teaches us in all things*."[188]

For no reason, then, does someone flatter himself for his tenacious memory, his persistent zeal, or his fluent speech, since teaching belongs to God alone, who is the Lord of all knowledge.[189] Consequently, in our prayers we should ask for knowledge from God, James 1: "If anyone lacks wisdom, let him ask God for it."[190] And Augustine: "Matters in doubt are better resolved by prayer than by any investigation."[191] And Bernard to the brothers of Mont Dieu: "We should summon up the desire [for prayer] from the course of the readings and compose a prayer that may interrupt a reading, but still not hinder it."[192] And Wisdom 7: "I prayed, and meaning was given to me, and I entreated, and the spirit of wisdom came to me."[193] And therefore one reads that when the blessed Bernard* wanted to study, he used to open a book and say this prayer first: "Depart from me, evil ones, and I will investigate the commandments of my

184 At the end of *De magistro*, Augustine's long monologue to his son Adeodatus provides the basis for Peter's summary here. Augustine describes learning as a process of grasping what is signified by signs that is an internal phenomenon for each student (not a transfer of information from teacher to student) guided by the presence of Truth (that is, God) within the soul. See Augustine of Hippo, *De magistro*, 11.37–14.45, ed. Klaus-Detlef Daur, in *Contra academicos; De beata vita; De ordine*, CCSL 29 (Turnhout, 1970), 194–202.

185 Rev 3:7.

186 Jerome, *Ep.*, 58.9.2, ed. Hilberg, CSEL 54:539.

187 Bernard of Clairvaux, *Epistolae*, 108.2, ed. Jean Leclercq and Henri M. Rochais, in *Opera* (Rome, 1974), vol. 7, 278.

188 1John 2:27.

189 Cf. 1Sam 2:3.

190 Jas 1:5.

191 Peter the Chanter (Petrus Cantor), *Verbum abbreviatum* (textus conflatus), 2.34, ed. Monique Boutry, CCCM 196 (Turnhout, 2004), 734.

192 William of Saint-Thierry, *Ep. ad fratres de Monte Dei*, 123, ed. Honemann, 336 (cf. PL 184:328A).

193 Wis 7:7.

God."[194] For the study of the wisdom that brings salvation is most displeasing to unclean spirits. And as Hugh of Saint-Victor says in *On Sacraments*: "We have a triple eye, namely an eye of contemplation, an eye of reason, and an eye of the flesh."[195] Because of the sin of the first man, our eye of contemplation has been extinguished, our eye of reason has become inflamed, but the eye of our flesh has remained clear-sighted. Thus, when a person thinks about the blindness and indistinctness of his intelligence and perceives that he can know nothing without the illumination of divine light, what remains for him to do besides saying in a prayer, "God, my God, shine light on my darkness"?[196]

194 Ps 118:115. Cf. Bernard of Clairvaux, *Ep.*, 11.1, ed. Leclercq and Rochais, in *Opera*, vol. 7, 53.

195 Hugh of Saint-Victor, *Commentariorum in Hierarchiam Coelestem s. Dionysii Areopagitae libri x*, 3, PL 175:976A.

196 Ps 17:29.

CHAPTER TWELVE

On the Instruction of Prelates Based on Seven Properties of the Eyes

The eyes are located in a very prominent part of the body; they keep watch for the entire body; they contemplate things that are distant from it, not those very close to it; they become frail and weakened in an aging body; they weep as if showing sympathy with the pains in other bodily members; they have eyelids with hairs, the size of which does not increase; and an injury that in other bodily members is thought to be moderate is considered serious when it happens to the eyes.

The first instruction for prelates. That just as prelates surpass others in their position, so should they in holiness.*

First, then, the eyes occupy a position in the body that is superior to other locations, and this is so because of the greater honor they have among the other bodily members. In this fact nature acts as a teacher to instruct the spiritual eyes of the church, that is to say, prelates, who have been given to the church to act as its eyes: Just as they surpass others in the honor of their position, so should they also rise above others in the holiness of their life. For as Gregory says: "A person who has been set over others as someone worthy of being imitated should behave in a distinguished manner."[1] And Bernard to Eugene: "Listen to a song less pleasant, but wholesome! It is a monstrous thing to be highest in rank but lowest in mind, first in the seating order but last in one's way of life, with an eloquent tongue but lazy hands, great when it comes to talking but devoid of results."[2] And for this reason one reads that Emperor Constantine wrote back as follows to his mother Helen when she had commanded him to do something unjust: "The higher our place among the people, the more we are exposed to their sight. And thus the eyes of all mortals carefully examine the judgments all of us make and our promises,

1 Magnus Aurelius Cassiodorus, *Var.*, 4.3.1, ed. Fridh, CCSL 96:144–45.

2 Bernard of Clairvaux, *Tract. de consid.*, 2.7.14, ed. Leclercq and Rochais, in *Opera*, vol. 3, 422.

whether we like it or not. And for that reason what we want must be not just blameless, but even praiseworthy."[3]

A large rock falling down a mountain carries along many rocks underneath it and causes them to fall. So also when a prelate has sinned, he causes many others to sink into the abyss of hell with him, as Gregory says: "When a shepherd tumbles down the steep cliffs of sin, inevitably his flock will be led to a headlong fall."[4] And since prelates have been given a high position, they can and they should rightfully fear a fall. For this reason the blessed Bernard says the following in one of the letters he writes to Eugene, who was newly elected pope: "From my son, Bernard, you* have been made my father, Eugene. Truly, this is an enormous change: You who came after me have been placed before me. I confess, I am not envious. For I think of your high degree, and I fear your fall; I think of the summit of your honor, and I tremble at the danger lurking nearby because of what is written: 'When a man is honored, he forgets what he is,'[5] as if it were saying: 'Honor has swallowed up his understanding.'"[6] A prelate must consider that the higher his rank, the more serious his fall. For this reason one reads in 2 Kings 1: "Ahaziah fell through the lattice of the upper story he made in Samaria"[7] and he died. He would not have been injured so severely from the fall if he had not climbed up so high. These sentences by Donatus should ring very often in prelates' ears: "How many grammatical elements belong to the preposition? One. That one being? Case alone. How many cases? Two. Which ones? The accusative and the ablative."[8] For the prelate ought to fear these things: to be accused (*accusari*) of a crime and to be removed (*auferri*) from his leadership position, and thus to fall (*cadere*) disgracefully.[9] An eye located in a high position judges that things lower than itself are

3 Vincent of Beauvais, *Spec. hist.*, 13.50, in *Speculum quadruplex*, vol. 4, 522a.

4 Gregory the Great, *Reg. past.*, 1.2, PL 77:15C.

5 Ps 48:13.

6 Bernard of Clairvaux, *Ep.*, 238.1–3, ed. Leclercq and Rochais, in *Opera*, vol. 8, 116–17. Pope Eugene III's name at birth was Bernardo Pignatelli. Before his election as pope, he had been a disciple of Bernard of Clairvaux.

7 2Kgs 1:2.

8 Aelius Donatus, *Ars minor*, 8, ed. Louis Holtz, in *Donat et la tradition de l'enseignement grammatical. Étude sur l'Ars Donati et sa diffusion (IVe – IXe siècle) et édition critique*, Documents, Études et Répertoires publiés par l'Institut de Recherche et d'Histoire des Textes, 24 (Paris, 1981), 600.

9 An etymological play on words is involved here: *accusativus* (the accusative), from *accusare* (to accuse); *ablativus* (the ablative), from *auferre* (to take away, to remove). This passage is connected to the previous discussion of a physical and moral fall by two meanings of the word *casus*: both "fall" and "grammatical case."

small, and for this reason if a person examines things beneath him, he ought to be afraid lest he grow light-headed as a result and fall. So, too, while he examines his subjects from on high, a prelate ought to be very afraid lest he grow light-headed through pride and thus fall spiritually. Therefore, it is said in Ecclesiasticus 32: "Have they made you a leader? Do not be exalted; among them, be like one of them."[10] And for this reason concerning someone newly elevated to be king it is said in Deuteronomy 17: "His heart should not be raised up in pride over his brothers."[11]

Moreover, from the fact that in the human body the eyes are located higher than the mouth and also that the eyes are twins while the mouth is one of a kind, the prelate is taught spiritually that he should focus his attention principally and fully more on circumspection than on speech – this is why the word "bishop" is interpreted to mean "watching over."[12] In the same way, the singularity of the mouth and the duality of the hands also warn him that he should pay more ample attention to deeds than to words. For it is fitting that a secular lord and a prelate be short on words and long on deeds. And therefore Job, who says, "And while I was sitting as a king, with the army surrounding me,"[13] says, "My word fell in drops upon them," Job 29.[14] But there are many prelates who are more capable of saying, "And while I sat around in public, with my companions surrounding me, my word swamped them." But let them consider that in loquacity sin is not wanting, and that the word of a prelate is judged priceless when he says little but does so wisely. For this reason Job also says in the same passage, "They did not dare add anything to my speech, and my word fell in drops upon them."[15] And for this reason Aristotle says as follows in *On the Secrets of Secrets* which he wrote to Alexander: "It is splendid and honorable for a king to refrain from loquacity unless it is necessary or reason demands it. For having people's ears always thirsty for his word is better than having them glutted by his speeches, since all overindulgence is loathsome."[16] Thus, if speaking only

10 Sir 32:1.

11 Deut 17:20.

12 Lat. *episcopus* (bishop) is derived from Gr. *episkopos* (overseer) = *epi* (upon) + *skopos* (one who watches).

13 Job 29:25.

14 Job 29:22.

15 Ibid.

16 Ps.-Aristotle, *Secretum secretorum*, 1.11, ed. Robert Steele, in *Opera hactenus inedita Rogeri Baconi* ... Pt. 5: *Secretum secretorum cum glossis et notulis ... accedunt versio Anglicana ex Arabico edita per A. S. Fulton, versio vetusta Anglo-Normanica nunc primum edita* (Oxford, 1920), 49. On this passage, see Steven J. Williams, *The Secret of Secrets. The Scholarly Career of a Pseudo-Aristotelian Text in the Latin Middle Ages* (Ann Arbor, 2003), 252n295.

a little is fitting for a king, how much more so for a pastor of souls? And therefore when someone asked of the philosopher Socrates "how a person could procure the best reputation, he responded, 'If he performs the best deeds and speaks few words.'"[17]

The second instruction for prelates. That it is the duty of a watchman, that is to say a prelate, to keep watch and not to be blind.*

Second, the eyes keep watch for the entire body. And in the same way, a prelate has to fulfill the office of a watchman in the church, Ezekiel 33: "Son of man, I have given you as a watchman to the house of Israel."[18] It is a watchman's duty to stay awake, lest if he fall asleep from bodily fatigue the enemy's sword strike those who are unguarded. But let the ones keeping watch over us hear what Bernard would say to them, in *Sermon 77* on the Song of Songs*, on the verse, "The watchmen found me, those who were guarding the city":[19] "It is not enough," he says, "that those watching over us neglect to keep guard, but they destroy us as well. Indeed, plunged in the deep sleep of oblivion, they do not even awaken at the thundering of divine threats to grow terrified at their own danger. So it is that those who do not take pains for themselves take no pains for their flock, destroying and at the same time being destroyed. Who can you name among all our leaders who is not more vigilant about emptying the purses of those in his care than about uprooting their vices? Oh, if only they were found to be as vigilant in caring [for souls] as they are swift in running for the position of bishop! Surely they would keep watch then, and they would not let it be said of them, 'My friends and my neighbors have drawn near, and stood against me.'"[20] If only our watchmen were at least as careful sentinels as cranes! For as the *Physiologus* says, and Hugh of Saint-Victor in the *Book of Judges*: "The cranes that are chosen to watch over the others hold a pebble in the foot they raise from the earth, so that if by chance one of them dozes off, the pebble will fall from its foot and then, awakening, it will cry out. Thus, the prelate who keeps watch to guard himself and others should carry a pebble in his foot, that is to say Christ in his mind: for the stone is Christ; the foot is the inclination of the mind. Above all things let him beware lest the pebble

17 John of Salisbury, *Policr.*, 5.6, ed. Webb, vol. 1, 304–05.

18 Ezek 33:7.

19 Cant 3:3.

20 Ps 37:12; Bernard of Clairvaux, *Serm. sup. Cant.*, 77.1.1–1.2, ed. Leclercq, Talbot, and Rochais, in *Opera*, vol. 2, 262–63.

slip from his foot, i.e., Christ from his mind. If it falls, however, let him call out through confession so that he awakens the sleepers, i.e., so that he summons the populace committed to his care to be vigilantly alert as much for himself as for their own lapses."[21] Therefore, it is said in Proverbs 6: "Son, if you have taken a pledge for your friend, run, hasten, rouse your friend. Do not let sleep come to your eyes, do not let your eyelids slumber."[22]

But it is said in Isaiah 56: "All his watchmen are blind."[23] For they are like Eli who was blind: They remain behind in the doorway and merely inform themselves about the Lord's army, not wanting* to undergo the hard work of being in the army nor go into battle with the soldiers.[24] They are also like another Jacob, who was blind and after crossing his hands placed his right one on Ephraim, since they focus their primary intentions and their deeds on worldly things, which are signified by Ephraim whose name means "fruit-bearing."[25] He placed his left hand on Manasseh, whose name means "oblivion"[26] and who therefore signifies heavenly things – which have been consigned to oblivion by many since these kinds of people fix their secondary intentions and deeds on heavenly matters.[27] For they mainly carry out actions with the right hand, but rarely, and only secondarily, with the left one. It is also obvious that they are blind, for they are like another Lamech, who killed a human being when he thought he was shooting an arrow at a wild beast.[28] And likewise, in our modern age, although they often seem to be in pursuit of someone's sin, they are in fact pursuing his rank. According to the Hebrew tradition, one reads that this Lamech was the first to suffer dimsightedness and to practice bigamy for the first time.[29] And it is definite and certain that spiritual bigamy in the church has its origin in the blindness of its prelates. One reads in 2 Samuel 5 that the Jebusites said to

21 Hugh of Fouilloy (Ps.-Hugh of Saint-Victor), *Aviarium*, 2.44, ed. and trans. Clark, 204–05.

22 Prov 6:1; Prov 6:3–4.

23 Isa 56:10.

24 Cf. 1Sam 4:15–16.

25 Jerome, *Lib. interp. Hebr. nom.*, Gen., ed. de Lagarde, CCSL 72:65. The Hebrew *ephraim* is derived from *parah* (to bear fruit) in Gen 41:52, though the word is more likely derived from *apar* (enclose, cover).

26 Jerome, *Lib. interp. Hebr. nom.*, Num., ed. de Lagarde, CCSL 72:82. On the etymology of the Hebrew *menasheh* from the root *nashah* (to forget) in Gen 41:51, see above, chapt. 6.10, p. 35.

27 Cf. Gen 48:14–20.

28 Cf. *Gloss. ord.*, Gen. 4:23, ed. Rusch, vol. 1, 33a in margin; cf. Gen 4:23.

29 Ibid., Gen. 4:19; cf. Gen 4:19.

David, "'You will not come in here, unless you remove the blind and the lame' who say, 'David will not come in here.'"[30] And there follows: "The blind and the lame will not enter the temple."[31] They are blind in whom the eye of the intellect has been blinded by ignorance and the eye of desire by malice. Someone is called lame when he has obtained two benefices in two churches and is unable to reside in both places at the same time and therefore appears to be limping spiritually, like a cripple with legs of unequal length. He is also blind, since he cannot see to both benefices. And these men are said to be located within the walls of Jerusalem and to forbid David from entering the temple, since today such people have commonly been placed in powerful spiritual offices and hinder good men from being promoted in the church. Thus, since they are blind, they are not able to say the verse truthfully in Job 29: "I was the eye for the blind and the foot for the lame."[32] And Bernard says: "A blind sentinel, an ignorant professor, a mute herald – these are monstrous things."[33] Their blindness is clearly evident, since we see them stumbling on the path of good behavior more than the laity. And therefore those who ought to guide others by acting as their eyes cause them to turn aside from the right course. Moreover, they are called blind for this reason: that even though they have eyes, they still do not use them for their own guidance, for frequently they are knowledgeable, but they do not lead their lives in accordance with their knowledge. And in this way they are comparable to the peacock that has many eyes on its tail, but sees nothing with those eyes.[34]

And therefore it often happens that prelates and many others who have been enlightened by knowledge do not taste the sweetness of God.

30 2Sam 5:6.

31 2Sam 5:8.

32 Job 29:15.

33 Cf. William Durand of Mende, *Ratio. div. off.*, 1.7.15, ed. Davril and Thibodeau, CCCM 140:88; cf. Gilbert of Tournai, *Collectio de scandalis ecclesiae*, 7.2, ed. Autbertus Stroick, in "Collectio de scandalis ecclesiae. Nova Editio," *Archivum Franciscanum Historicum* 24 (1931): 33–62, here 41; Iacopo da Varazze (Jacobus de Voragine), *Legenda aurea*, 178.113–14 ("De dedicatione ecclesie"), ed. Giovanni Paolo Maggioni, in *Legenda aurea. Con le miniature del codice Ambrosiano C 240 inf.*, Italian trans. Gianfranco Agosti et al., Edizione nazionale dei testi mediolatini, 20; Serie II, 9 (Florence and Milan, 2007), vol. 2, 1446–48. Iacopo is quoting "Gilibertus."

34 Note that in Hugh of Fouilloy (Ps.-Hugh of Saint-Victor), *Aviarium*, 2.59, ed. and trans. Clark, 248–49, the eyes on the peacock's tail are not part of a critique of teachers, but rather are said to signify that "every teacher can foresee the danger threatening each of us in the end."

What Elisha said to someone in 2 Kings 7 can be said to them: "You will see it with your own eyes, and yet you will not eat of it."[35] But the illiterate and simple-minded who do not have eyes for knowledge frequently experience and taste the sweetness of divine contemplation and consolation. And they can be designated by Isaac, whose "eyes were blinded so that he could not see"[36] when he ate the savory food brought to him by his son, as it is said in Genesis 27.[37] As Perspectivist science maintains, nature doubled the eyes to two for this reason: so that with one eye closed or even obstructed, the visual spirit[38] could transfer itself to the other one and operate more powerfully in it. For then the pupil of the eye is strengthened in the act of seeing, since the power of sight is sustained in it more abundantly.[39] We see this in archers, who close one eye so they can shoot their arrows more exactly. And likewise, although we see God at the present moment with doubled eyes, that is to say, those of reason and faith, sometimes when the eye of reason is closed, the eye of faith contemplates God more clearly and more securely. For this reason it happens occasionally that a simple old woman will most faithfully grasp an article of the faith that an educated man, relying on reason, will be uncertain about. One reads in Genesis 48: "The eyes of Israel grew blind with advanced old age and he could not see clearly."[40] Israel is interpreted to mean "seeing God,"[41] and it signifies those who are educated and who search for God in the Scriptures, but who nevertheless are blind with respect to the deeds they should do. On the other hand, about Moses one reads in Deuteronomy 34* that "his eye did not grow blind."[42] Moses, whose name is interpreted to mean "drawn from the waters,"[43] signifies the simple-minded and uneducated who have been separated from the waters of wisdom through ignorance, but in matters which pertain to ethical behavior these people frequently see more clearly. For with their

35 2Kgs 7:2; 2Kgs 7:19.

36 Gen 27:1.

37 Cf. Gen 27:17.

38 On the visual spirit(s), see above, chapt. 7.5, p. 54, n50.

39 Peter discusses the doubling of the eyes above in chapt. 7.1, p. 45. Neither Roger Bacon nor Alhacen mentions that one eye is strengthened when the other one is injured.

40 Gen 48:10.

41 Jerome, *Lib. interp. Hebr. nom.*, Ex., ed. de Lagarde, CCSL 72:75. The etymology is incorrect; the Hebrew *yisrael* is derived from *sarah* (to persist, contend) + *el* (God) in Gen 32:29, which is linguistically possible.

42 Deut 34:7.

43 Jerome, *Lib. interp. Hebr. nom.*, Ex., ed. de Lagarde, CCSL 72:76. The Hebrew *moshe* is derived from *mashah* (to draw out) in Exod 2:10; however, the etymology of the name remains unclear.

excessive research the educated are blinded by an excess of the objects of sense. For according to Aristotle in book 2 of *On the Soul*: "Excesses of the sensibles disrupt the senses."[44] And for this reason concerning Paul one reads in Acts 9: "Suddenly a light from heaven flashed around him, and when he opened his eyes, he saw nothing."[45]

The third instruction for prelates. That a prelate should carefully scrutinize not only others' behavior, but also the behavior of those close to him.*

Third, the eyes do not see things that are very close to them, but things that are at some remove. So, too, prelates scrutinize other people's lives, but they do not investigate nor do they pay attention to the behavior of their friends and their own household, so that they can truly say this verse in the Psalm: "The light of my eyes – it is no longer with me."[46] For as Jerome says in *Letter 46 to Sabinian**: "Generally, we are the last to know the scandals affecting our own house, and we remain ignorant of the vices of our children and our spouses, though the whole neighborhood knows about them."[47] If the eye [of these prelates] were turned back on itself, so that they would examine their own life and the life and behavior of their friends and family, it would from time to time observe many evils in them. For "as a snare is full of birds, so are their houses filled with deceit," as it is said in Jeremiah 5.[48] "The sword kills outside; within the house it is like death itself," Lamentations 1.[49] Indeed, [their eye] would see* pride in their own house, according to the verse by Juvenal:

> Every aristocratic house is filled with proud servants.[50]

44 Aristotle, *De anima*, 3.4 (429b2), ed. Ross; trans. Smith (see also Thomas Aquinas, *Summa theologiae*, Ia, quest. 88, art. 1, argumentum 3, in the Blackfriars ed. [London and New York, 1968], vol. 12, 120–21, with the exact Latin text that Peter gives). Aristotle's point is that when a sense perception is too strong, it disrupts the normal pattern of sensation; for example, a sound that is too loud leaves us unable to hear for a while.

45 Acts 9:3; Acts 9:8.

46 Ps 37:11.

47 Jerome, *Ep.*, 147.10.3, ed. Hilberg, CSEL 56:327.

48 Jer 5:27.

49 Lam 1:20.

50 Juvenal, *Sat.*, 5.66, ed. Clausen, 68 (cf. Walther, *Proverbia*, no. 14509).

But David says, "No one who commits deeds of pride will dwell in the midst of my house."[51] Such people ought to be ejected and removed from the house of prelates, no matter how much they seem otherwise to belong there, according to the verse in Matthew 5: "If your eye scandalizes you, pluck it out and throw it away from you."[52] For often when such a person is plucked out, the safety and health of many are ensured by the expulsion of one. When those who are left reflect on this, they examine themselves and say, "Someone who vents his rage in front of our eyes – who would he spare? Therefore, let's correct our own behavior."

> For your own property is affected when the neighbor's wall is in flames.[53]
> Happy is the man made wary by another's peril.[54]

And if a prelate would carefully examine his own house, he would occasionally find in it people with mutual hate for each other and in a state of mutual envy. For as the common expression has it: "Those dwelling at court are like dogs in the kitchen."[55] For this reason:

> While a dog gnaws a bone,[56] and so on.

And as Claudian says:

> The palace servants haughtily despised his company.[57]

A dog does not eat the plants in a meadow, and yet it does not allow* other animals to eat them. So, too, one sometimes finds people at the

51 Ps 100:7.

52 Matt 5:29, cf. also Matt 18:9.

53 Horace, *Ep.*, 1.18.84, ed. Wickham, 2nd ed., ed. Garrod; cf. Walther, *Proverbia*, no. 15871.

54 *Sententia communis* (Walther, *Proverbia*, no. 8952).

55 The saying is proverbial. See the English version: "as hende as hound is in kitchen," used in *Piers Plowman*; cf. Bartlett Jere Whiting and Helen Westcott Whiting, *Proverbs, Sentences, and Proverbial Phrases from English Writings Mainly before 1500* (Cambridge, MA, 1968), H558.

56 *Sententia communis* (Walther, *Proverbia*, no. 6445). The complete proverbial expression is: "Dum canis os rodit, socium, quem diligit, odit" (While a dog gnaws a bone, it hates its friend, whom it otherwise loves).

57 Claudius Claudianus, *In Eutropium*, 1.149–50, ed. John Barrie Hall, in *Carmina* (Leipzig, 1985), 149. The point Claudian is making is that even the palace servants scorned Eutropius, a eunuch and formerly below them in station, who rose to power as a consul in the Eastern empire in 399 CE.

courts of prelates who are unworthy of benefices, and nevertheless they stop worthy people from being admitted to one. The prelate would also see in his own house people who have given themselves to lust, so that he would be able to say that he dwells in a house of mud, Job 4.[58] But David says, "He who walks on a blameless path, he will serve me."[59] And about the blessed Edmund*, archbishop of Canterbury, one reads that "he demanded cleanliness of the flesh from his servants, wanting everyone in his household to be clean at all times."[60] He will also see wolves in sheep's clothing in his own house endeavoring to lead astray the master of the household. Bede says that the serpent which led Eve astray had a human face and the tail of a serpent.[61] The evil advisors who are frequently found in the house of a prelate are similar to this, for their speech appears to be rational and level headed, but its end is daubed with poison. Thus, give heed, father and master of the household, lest you be deceived by:

Evil intentions* concealed under a fox pelt.[62]

For perhaps subverters are under your roof, "and you are dwelling with scorpions," Ezekiel 2.[63] He will also see "acts of injustice in the hands" of his companions, for "their right hands are filled with bribes."[64] And in the end, when he sums up everything in one [statement], he will also be able to see that "his house is not without someone suffering a discharge of semen, and one who is leprous, and one holding a spindle, and one felled by the sword, and one lacking bread," 2 Samuel 3.[65] For he will see the discharge of semen in the lustful, leprosy in disparagers, a sword in the wrathful and the envious, and the lack of bread in the greedy and the avaricious. When he has seen these things, he will truly be able to say that

58 Cf. Job 4:19.

59 Ps 100:6.

60 *Vita b. Edmundi*, 27, ed. Martène and Durand, in *Thesaurus novus anecdotorum*, vol. 3, 1796A.

61 Cf. the similar language, though not attributed to Bede, in Thomas of Chobham, *Summa de commendatione virtutum et extirpatione vitiorum*, 1, ed. Franco Morenzoni, CCCM 82B (Turnhout, 1997), 49. Peter Comestor, *Scholastica historia: Liber Genesis*, 22, ed. Agneta Sylwan, CCCM 191 (Turnhout, 2005), 40, attributes to Bede the opinion that the devil had the shape of a serpent and the face of a virgin when he tempted Eve. See also Payer, *The Bridling of Desire*, 206n1.

62 Cf. Horace, *De arte poetica*, 437, ed. Edward C. Wickham, 2nd ed., ed. Heathcote W. Garrod, in *Opera* (Oxford, 1901).

63 Ezek 2:6.

64 Ps 25:10.

65 2Sam 3:29.

passage in Jeremiah 23: "I have found evil in my own house."[66] And with grief, and at the same time with surprise, he will be able to exclaim and say, "Has this house, then, been made a den of robbers?" – Jeremiah 7.[67] Looking at the person in his household whom he favors most, he will also exclaim with surprise, "Why is it that the one I favor has committed many vile deeds in my house?"[68] He will also see that what Bernard says to Eugene is true: "The Curia usually accepts good men more easily than it produces them."[69] And again in the same work: "I would not say that you were healthy if your sides ache."[70] Therefore, it is taught in Levitcus 13 and 14 that the priest should investigate and inspect not only the leprosy which infects human beings,[71] but also the one that eats away at houses, since a prelate should not only examine the life of others, but also pay careful attention to the behavior of their friends and family. And after he has inquired into the merits of strangers, he should return to himself and should say with the patriarch Jacob, "It is reasonable that sometimes I should look after my own house as well," Genesis 30.[72]

It is very fitting in all ways for a prelate to maintain a respectable household, for as Seneca says: "A well-ordered prince ought to have a well-ordered household."[73] And Boethius, in *The Consolation of Philosophy*: "Servants," he says, "of immoral character are a destructive burden for a house, and they are an extremely injurious one to their lord."[74] And Juvenal the satirist:

> There are many reasons for living correctly, yet among them
> This above all, that you scorn the chatter of your
> Domestics. For the tongue is the worst part of a bad servant.[75]

66 Jer 23:11.

67 Jer 7:11.

68 Jer 11:15.

69 Bernard of Clairvaux, *Tract. de consid.*, 4.4.11, ed. Leclercq and Rochais, in *Opera*, vol. 3, 457. Bernard speaks here of the Papal court.

70 Ibid., 4.4.9, vol. 3, 455.

71 Cf. Lev 13:1–14:57.

72 Gen 30:30.

73 Cf. Ps.-Albert the Great (Adenulf of Anagnia [?]), *Commentarii in Psalmos*, Ps. 100.7, in *Opera omnia*, ed. Auguste Borgnet (Paris, 1892), vol. 16, 598. Ps.-Albert also admonishes prelates to have a respectable household in this passage, likewise attributing to Seneca the sentiment that "a well-ordered soldier (*miles*) should maintain a well-ordered and chaste household."

74 Boethius, *Phil. cons.*, 2.pr.5.18, ed. Bieler, CCSL 94:27.

75 Juvenal, *Sat.*, 9.118–21, ed. Clausen, 120 (cf. Walther, *Proverbia*, no. 33964b).

For we see in the physical world that the sun cannot illuminate things that are remote from it without first illuminating things that are in between itself and those remote bodies or things that are close to it, nor can the sun illuminate things that are next to it without shining on its own substance. And in the same way, a prelate, who is like the sun in the church, cannot very well illuminate the church given to his care without illuminating his own household as well. For as it is said in 1 Timothy 3: "If someone does not know how to rule his own household, how can he exhibit care for God's church?"[76] For if the ray of light is darkened in a neighbor, i.e., through improper associates and advisors, it will inevitably reach the others. Nor can a prelate have a good household without also leading a good life himself. For a pupil wants to be like his instructor, as it is said in Ecclesiasticus 10: "Like a people's judge, so also their minister; and like a city's governor, so also its inhabitants."[77]

*The fourth instruction for prelates. That prelates and clergymen, whose eyes are weakened, are worse than the laity**.

Fourth, the eyes become frail and weakened in an aging body. So, alas!, as if the church is becoming old, we see that its eyes have grown weak now, while as time progresses the prelates and clerics become worse than the laity, so that the prophecy of Isaiah 24 has been fulfilled today: "And it will be that as the people are, so also will the priest be."[78] "And if only priests would not be worse than the people," as Bernard adds.[79] One reads in Genesis 27: "Isaac was old and his eyes grew blind and he could not see."[80] And since Jacob presented the savory food to him and he ate it, and thus his sense of taste was not deceived, neither was his sense of touch that very clearly felt the hairiness of Jacob's neck and hands; nor his ears, for he said, "The voice is the voice of Jacob";[81] nor his sense of smell, since "he smelled the fragrance of Jacob's garments."[82] Only his sight failed him.

76 1Tim 3:5.

77 Sir 10:2.

78 Isa 24:2.

79 Cf. Peter the Chanter (Petrus Cantor), *Verb. abbrev.* (text. confl.), 1.55, ed. Boutry, CCCM 196:375: "Et erit sicut populus, sic sacerdos, immo stultior est hodie quam populus sacerdos" (And it will be that as the people are, so will the priest be, but today the priests are more foolish than the people).

80 Gen 27:1.

81 Gen 27:22; cf. Gen 27:17–27.

82 Gen 27:27.

So, too, in this waning twilight of the world we see that the prelates of the church have fallen away from their original state of perfection more than the laity has. And therefore the state of the church can be symbolized by the statue which Nebuchadnezzar saw in his dreams, as one reads in Daniel 2: "The head of this statue was made of gold, the chest and arms of silver, the belly and thighs of bronze, but the feet were partly of iron and partly of clay."[83] For the church was formerly golden in wisdom in its past prelates, like silver in the purity of their life, and like bronze in their eloquence – three things that are necessary for a preacher, namely, the gleam of his wisdom, the purity of his life, and the melodiousness of his eloquence. But the feet, i.e., the church's most recent prelates, namely our modern ones, are partly iron, due to their hard-heartedness, and partly clay, due to their fleshly lust. When Archbishop Peter of Sens* of blessed memory once came to the abbey at Clairvaux to* meet with someone in that house who was frequently transported in ecstasy and to consult with the man, he requested that he relate one of his visions. "Recently," the man said, "I had a vision of a lady with the most beautiful visage, adorned with gold and jewels, and as I was astonished by her adornment and was marveling at the beauty of her visage, she said to me, 'Just who am I?' And I responded, 'It seems to me you're the Blessed Virgin.' And she said, 'You should look at me from behind.' And when I looked at her back, I saw that she was decayed and crawling with worms. 'Now,' she said, 'you can clearly comprehend that I'm not the glorious Virgin. For I'm not the Virgin Mary, but Mother Church. In my initial state, and the part in front, I was most beautiful and nobly adorned with the apostles, martyrs, and confessors, but from behind, that is to say in the final days, as in modern prelates, I'm decayed and corrupt and full of disgrace, and "my silver has been turned into slag," Isaiah 1.'"[84] Valerius Maximus relates in book 6 that when everyone was hoping for the death of Dionysius as the cruelest tyrant, a certain woman of very advanced age was living at Syracuse who alone used to pray to God most devoutly every day during the morning hours that Dionysius remain alive for a long time. As soon as he learned of this fact, Dionysius, who was astonished by the woman's benevolence towards him, immediately summoned her to come to him and asked about the reason for this and how he had deserved it. "When I was a girl," she said, "we had a cruel tyrant as lord and

83 Dan 2:32–33.

84 Isa 1:22. For the narrative, cf. Tubach, *Index exemplorum*, nos. 1051 (concerning the personification of Lady Church) and 5390 (concerning the personification of Lady World). Peter of Corbeil (d. 1222) was the teacher of the future Pope Innocent III, who then promoted Peter to be archbishop of Sens in 1200.

we used to yearn devoutly to be rid of him. After he died, a crueler one succeeded him, and his death was an even greater desire for us than that of the previous one. Finally, we came to have the third of our leaders, namely you, more dreadful than your predecessors. Thus, it has been my custom to pray to God daily to let you have a long life and for this reason: lest someone who is the worst of all become your successor if death takes you away."[85]

The church, alas!, is developing in this way if one considers its earlier days and its present ones. For formerly Moses used to cover his face, since "the children of Israel could not look at him because of the glory of his face," as is said in 2 Corinthians 3.[86] But today our Moseses, i.e., the priests of modern times, can cover their face out of embarrassment. For now the children of Israel not only refuse to pay attention to them, but rather blush with shame and groan with chagrin. One reads in 2 Maccabees 3: "But now whoever saw the high priest's face was pierced to the heart."[87] But today, if one thinks carefully about the state of the church, one is pierced to the heart. Considering this disparity among prelates of the church, Gregory says: "On the day of judgment it will be shown what each and every one has won for the church. There, Peter will appear with a converted Judea; there Paul with the Gentiles converted to the faith; there Andrew leading Greece after him; John, Asia; Thomas, India. Thus, since so many shepherds will come before the judge with their flocks, what will we miserable ones say who return to the Lord empty-handed after our labor, we who have the name* of shepherds and yet we have no sheep that we have an obligation to show as the result of our nurturing? Here we are called shepherds, but there we have no flock to lead."[88]

The fifth instruction for prelates, on prelates' dutiful compassion for their congregants, and on the ostrich egg.*

Fifth, the eyes weep as if they had compassion with the pains of other bodily members. So, too, when a prelate has examined the misfortunes of his congregants, he should burst out in tears of compassion, on the example of what Jeremiah said in Jeremiah 13: "My eye will weep bitterly and will shed tears, since the Lord's flock has been taken captive."[89] The

85 Valerius Maximus, *Fact. et dict. mem.*, 6.2.Ext.2, ed. Briscoe, vol. 1, 386–87 (cf. Tubach, *Index exemplorum*, no. 1678).

86 2Cor 3:7.

87 2Mac 3:16.

88 Gregory the Great, *Hom. in Ev.*, 1.17.17, PL 76:1148B–C.

89 Jer 13:17.

blessed Job was this kind of person; he used to say, "I wept formerly for him who was afflicted, and my soul had compassion with the pauper."[90] Valerius Maximus relates in book 5* that after the capture of Syracuse, Marcellus stood on the city's fortress so that from there he could look down on the city which had formerly been so opulent and was now in ruins. Contemplating its pitiable misfortune, he could not restrain his tears; if someone unfamiliar with the situation had seen him, he would have thought that the other side had gained the victory. And so, Syracusan city, in your greatest defeat you had at least this consolation: that even if you were not able to remain unscathed, at least you fell to a most merciful victor.[91] If, therefore, this ruler was incapable of containing his tears at the sight of his enemies' defeat, how great should the river of tears be that flow from prelates' eyes when they see the spiritual defeat of souls entrusted to their guidance! Let them hear what Jeremiah says: "Who will give my head water and my eyes a fountain of tears, so that I may weep day and night for those who were slain?"[92] and so on. For they ought to weep with those who are weeping and through compassion be enfeebled with those who are sick, on the example of Paul who says in 2 Corinthians 11*: "Who is enfeebled, and I do not become sick?"[93] and so on.

It is the nature of the fish that when it sees danger, it takes little fish into its mouth until it has brought them to a safe location.[94] And likewise, a prelate who sees that one of his congregants is in danger should take him in his mouth through salutary prayer and a word of encouragement, and into the innermost chambers of his heart through compassion. Paul used to do this, who said in Philippians 1: "God is my witness how I yearn for you in the innermost flesh of Jesus Christ."[95] As Julius

90 Job 30:25.

91 Cf. Valerius Maximus, *Fact. et dict. mem.*, 5.1.4, ed. Briscoe, vol. 1, 298.

92 Jer 9:1.

93 2Cor 11:29.

94 The conception of fish protecting their young by taking them into their mouths is found in the bestiary tradition. See, for example, *Bestiarium*, 119, ed. Willene B. Clark, in *A Medieval Book of Beasts. The Second-Family Bestiary* (Woodbridge, 2006), 208. The passage in the *Bestiarium* is taken from Ambrose of Milan, *Exameron*, 5.3.7, ed. Karl Schenkl, *Opera*, pt. 1, CSEL 32.1 (Vienna, etc., 1897), 145–46. For pointing me to this material, I am indebted to Susan Crane. Cf. the related idea that whales swallow their young to keep them safe from storms at sea in Bartholomeus Anglicus, *De proprietatibus rerum*, 13.26 (Nuremberg, 1492), sig. q1rb. For an early translation into English, see the version by John Trevisa, ed. M.C. Seymour et al., in *On the Properties of Things. John Trevisa's Translation of Bartholomaeus Anglicus, De Proprietatibus Rerum. A Critical Text* (Oxford, 1975), vol. 1, 686.

95 Phil 1:8.

Frontinus* relates and also Valerius Maximus in book 5, when Codrus, king of the Athenians, was battling with the Dorians and had gathered together his army, he consulted the oracle of Apollo concerning the outcome of the war. The response to him was that his side would not gain victory unless he were killed by the enemy's sword, something which became known throughout the camp of the Athenians as well as among their opponents*, the Dorians. At that point, the Athenians entrusted their king's protection to the soldiers, but the Dorians decreed that no one was to harm Codrus' person. After Codrus heard this, he changed his royal attire, put on the clothing of a servant, and carrying brushwood* on his back went behind the enemy's lines. And one of their soldiers, whom Codrus hit with a scythe, hit him back, killing him.[96] Oh, for the welfare of his people how much this man wept, who for that very reason desired to have an enemy's sword come down on his head! Let sluggish prelates blush in shame, therefore, who in the face of this man's example do not erect themselves as a defensive wall for the house of God,[97] whose eyes are so dry that they have not shed a single tear, being slow to feel compassion for a perishing people.

Today that Gospel parable in Luke 10 is verified in which it is said: When a priest and a Levite saw a wounded man, they passed him by.[98] For we see both among the greater and among the lesser prelates that they have no compassion for souls who have been wounded spiritually, although inasmuch as they are spiritual physicians it is part of their duty to tend to the health of souls. For as Gregory says: "When a bishop-elect takes up pastoral care for the people, he approaches it like a physician does a sick person."[99] But as Bernard says to Eugene, speaking as if he were one of the [uncaring] kind of prelates: "We calculate back and forth our daily expenses with daily scrutiny, but we know nothing of the continual losses to the Lord's flock. There is a daily discussion with the servants about the price of food and the number of loaves of bread; rarely at

96 Cf. Valerius Maximus, *Fact. et dict. mem.*, 5.6.Ext.1, ed. Briscoe, vol. 1, 351; cf. the distant analogue in Sextus Julius Frontinus, *Strateg.*, 4.5.14, ed. Ireland, 104–05. The story is told in Justin (Marcus Junianus Justinus), *Epitoma historiarum Philippicarum Pompei Trogi*, 2.6.16–21, 2nd ed., ed. Otto Seel (Stuttgart, 1985), 25. Justin's name has apparently been mangled in transmission in the reference to Sextus Julius Frontinus. Cf. also John of Salisbury, *Policr.*, 4.3, in *Policraticus I–IV*, ed. Keats-Rohan, CCCM 118:238–39; Cicero, *Tusc. disp.*, 1.116, ed. Pohlenz, 277–78 (cf. Tubach, *Index exemplorum*, no. 1136).

97 Cf. Ezek 13:5.

98 Cf. Luke 10:31–32.

99 Gregory the Great, *Reg. past.*, 1.9, PL 77:22D.

all is a council called with the priests about the people's sins. A donkey falls, and someone will lift it up. A soul falls, and no one thinks about it."[100] For this reason a versifier says:

> Not for altars do they care, but for earnest moneys; not for true matters, but for coins;
> Not for what is equitable, but for a steed; not for the needy, but for the wealthy.[101]

It is surprising and deplorable that they grieve about the loss of a penny and are not sad about the loss of someone entrusted to their care.

It has sometimes been asked why ostrich eggs are hung in churches and what sort of people tell more lies. The first question is answered in the following way: Ostrich eggs are hung in churches and placed before prelates' eyes lest they imitate the ostrich in its neglect for its young.[102] For, concerning the ostrich it is said in Job 39: "It is callous with its young, as if they did not belong to it."[103] But the response to the second question has been as follows: Priests and clerics above all other people lie more frequently and, even worse, as a matter of custom. For daily they declaim before all, "A sense of failure seizes me because of the sinners who forsake your law."[104]

The sixth instruction for prelates. On the moderation prelates should preserve in temporal matters.*

Sixth, the eyes have eyelids with hairs, the size of which does not increase over time. Hairs, which are attached to animals on the outside, signify temporal matters that are added to a person as it were on the outside. For

100 Bernard of Clairvaux, *Tract. de consid.*, 4.6.20, ed. Leclercq and Rochais, in *Opera*, vol. 3, 464.

101 *Sententia communis* (Walther, *Proverbia*, no. 17433). The wordplay in this verse depends on a series of Latin homonyms, or near homonyms: *aras* (altars) and *aras* (= *arrhas*, earnest moneys), *vera* (true matters) and *era* (= *aera*, rich ores), *equum* (= *aequum*, equitable) and *equum* (steed), *inopes* (the needy) and *opes* (the wealthy).

102 On the use of ostrich eggs in churches, see Pierre Alain Mariaux, "Collecting (and Display)," in *A Companion to Medieval Art: Romanesque and Gothic in Northern Europe*, ed. Conrad Rudolph, Blackwell Companions to Art History, 2 (Malden, MA, 2006), 213–32, here 218–19, 221. Peter's text is further evidence that ostrich eggs were often suspended in the churches when put on display and that they were seen to have an admonitory function, to which Mariaux calls attention.

103 Job 39:16.

104 Ps 118:53.

this reason Matthew 6 reads: "All these things," that is to say, temporal matters, "will be added to you."[105] Thus, in the fact that the hairs on the eyelids do not grow over time but have a naturally determined size, the eyes of the church*, that is to say, prelates, are taught that they should not increase their temporal possessions, but should be content with a moderate amount. For if a worldly king is taught by the Lord in Deuteronomy 17 that he should not possess "immense amounts of silver and gold,"[106] how much more does he also say this to prelates and church officials, who are not lords but stewards of ecclesiastical wealth? But today these hairs have grown and increased to such a degree in prelates and other ecclesiastical personages (because they have been made hairy in every part of their bodies), that what was formerly predicted about Babylon in Isaiah 13 has now been fulfilled in the church: "There ostriches will dwell, and hairy satyrs will dance there."[107] The ostrich has wings but is unable to fly up in the air, being encumbered by the huge bulk of its body; it signifies hypocrites who counterfeit the life of good people on the outside, but because of their desires are never raised up from what is worldly to what is heavenly.[108] But the hairy satyrs are people who are wrapped up in temporal riches, and these people are said to dance in Babylon, i.e., in the church, in which nearly everything has been brought into disorder, since lewdness sometimes arises from worldly opulence, Deuteronomy 32: "The beloved one grew fat and became disobedient."[109] The same prophet is also mindful of these satyrs when he speaks about Babylon and says: "One satyr called to another."[110] For today we see in the church that a satyr invites a satyr, since a rich man entertains a rich man, although it is said in Matthew 14: "When you give a dinner or a feast, do not invite the rich, but invite the poor."[111]

This cursed hairiness has become very widespread among the prelates of the church, for "from the greatest to the least, all are eager for avarice"[112] and for storing up money. For they pay heed to what

105 Matt 6:33.

106 Deut 17:17.

107 Isa 13:21.

108 Cf. Hugh of Fouilloy (Ps.-Hugh of Saint-Victor), *Aviarium*, 2.42, ed. and trans. Clark, 188–89. The connection between ostriches and hypocrites is based on Gregory the Great: see *Mor. in Iob*, 7.28.36 and 31.8.11, ed. Adriaen, CCSL 143:360 and 143B:1557.

109 Deut 32:15.

110 Isa 34:14.

111 Cf. Luke 14:12–13.

112 Jer 6:13.

Cicero said: "Frugality makes the best income."[113] They make use of metaphors and *exempla* to foster avarice and cupidity. They say that if a pauper is frugal, he is most easily satisfied; however, if a rich man is extravagant, he is most easily cleaned out. For we see that even with small deposits every day, in the end the most spacious coffer will be filled up, and on the other hand, one that is full will be emptied out by repeated withdrawals. Through a small outlet for water, a pond is ultimately turned into dry land. We also see that if a channel flowing with water is diverted from a spring, not long afterwards the spring itself will run out of water and dry up. So, too, when payments are made from a heap of accumulated wealth for the most minute, but nevertheless still frequent and continuous expenses, in the end the riches will be totally exhausted by daily subtractions of these minute amounts. For as Claudian says:

> The plunderer of wealth is Riotous Living, to whom
> Unhappy Need always clings as a companion with humble steps.[114]

Therefore, one ought to be sparing in paying out money, but break out in a sweat amassing riches. And Juvenal says:

> Riches smell good, wherever they
> Come from. Let this saying always be on your lips,
> Worthy of the gods and of Jove himself as a poet:
> No one asks where someone's money comes from, so long as he has it.[115]

For according to the same poet:

> As much cash as a man has in his safe,
> Just so much faith do others have in him.[116]

And again:

113 Cicero, *De re publica*, 4.7, ed. Konrat Ziegler, 7th ed. (Leipzig, 1969), 110–11; idem, *Paradoxa*, 6.3.49, ed. Carl F.W. Mueller, in *Scripta quae manserunt omnia* (Leipzig, 1898), Pars 4, vol. 3, 212.

114 Claudius Claudianus, *In Rufinum*, 1.35–36, ed. John Barrie Hall, in *Carmina* (Leipzig, 1985), 14 (cf. Walther, *Proverbia*, no. 7589a).

115 Juvenal, *Sat.*, 14.204–07, ed. Clausen, 162 (cf. Walther, *Proverbia*, nos. 14003 [first three verses] and 32150 [last verse]).

116 Ibid., 3.143–44, ed. Clausen, 54 (cf. Walther, *Proverbia*, no. 23628).

Money has the victory, nor does he who newly came
To this city on whitened feet attain to the sacred office.[117]

With thoughts and enticements of this kind the love of possessing inflames these people's hearts. This is the vanity which Solomon testifies to having perceived under the sun in Ecclesiastes 4: "A man by himself without another," i.e., without an heir, "neither a son nor a brother, and nevertheless there is no end to his toil."[118] When he gathers up superfluous things, not knowing for whom he is amassing them, "his eyes are not satisfied with the riches"[119] he has acquired, but he always gapes* with longing at other things he wants to acquire. And in truth, that is vanity, or even more, great perversity, for as Aristotle says, no one thinks about applying the measure common for everyone to a goal, but rather to the means leading to a goal.[120] Thus, since good health is the goal for which a physician exerts himself, he does not consider the common measure promoting this state; he just promotes good health as far as he can. But to the means leading to this goal, that is to say the medication, he does take the commonly applied measure of it into consideration, since he does not administer medication in a merely random amount to the patient, in order not to kill him, but rather he determines the strength and the quantity of the medication as the disease and the strength of the

117 Ibid., 1.110–11, ed. Clausen, 40. Imported slaves put up for sale were marked with chalk on their feet. Peter's text (*Pr1*, with identical wording in *P7P8*: "Vincunt diuitie, nec sacro cedit honori / Nuper in hanc vrbem pedibus qui venerat albis") represents an interesting change from the text of Juvenal: "Vincant divitiae, sacro ne cedat honori / Nuper in hanc urbem pedibus qui venerat albis" (Let money have the victory, let he who newly came to this city on whitened feet not yield to the sacred office [i.e., the tribunes]). Both texts lament the power of money: Juvenal satirizes the attitude of the nouveau riche; Peter's verse emphasizes the necessity of money to achieve the priesthood.

118 Qoh 4:8.

119 Ibid.

120 Cf. Aristotle, *Eth. Nic.*, 3.5, trans. Robert Grosseteste (textus recognitus), ed. Gauthier, Fasc. 4, 416, where Aristotle notes: "Consiliamur autem non de finibus, set de hiis que ad fines" (We do not deliberate about ends, but about the means to ends), and goes on to use the example, among others, of the physician (acting *qua* physician) who does not deliberate about whether the goal is to cure a patient, but rather takes this for granted and considers which means will best achieve that goal. The passage in Aristotle's work is focused on the morally good deliberation that leads to choice, but for Peter by the end of this section the question becomes one of the limited appropriateness of temporal ends. I am grateful to Bonnie Kent for bringing the passage in the *Nicomachean Ethics* to my attention.

patient require. Now, we know that spiritual matters are related to a good end, and temporal matters lead to this end, as if it were their goal. Thus, spiritual goals ought to be promoted and desired without any question of the measure common to everyone, but temporal ones are fitting for each and every person only to a degree. Yet today we see the contrary of this taking place, for everyone applies the common measure to his own spiritual goals, saying, "It is enough for me to have this virtue, and to attain a particular level of such a virtue." But it is rare to find someone who applies the measure common for everyone to his own temporal goals and says, "This benefice is enough for me." But having obtained the first benefice, people seek a second one and after that a third and so on without limit. For:

> In someone who yearns, desire grows as his wealth grows.[121]

The seventh instruction for prelates: that sin is much more unsightly and harmful in a priest or a prelate than in a member of the laity.*

Seventh, what is considered a great injury to the eyes is occasionally thought to be only moderate when it happens to other bodily members. So, too, one and the same sin is greater in a prelate and a priest than in a member of the laity. And likewise a thorn causes a greater injury to the eye than to the foot or the hand. In the same way, it makes sense that there is just as much difference between the transgression of a prelate and one committed to his care, between a priest and a member of the laity, as there is between a mote and a beam, according to what is said in Matthew 7: "But why do you see the mote in your brother's eye, and you do not pay attention to the beam in your own eye?"[122] For just as a sin is all the worse when it is committed in a holy place, so, too, the status and rank of the person involved magnify his fault. And for this reason Juvenal says:

> The higher the status of the sinner,
> The more public the judgment of each fault in his soul.[123]

121 Cf. Juvenal, *Sat.*, 14.139, ed. Clausen, 160: "Crescit amor nummi, quantum ipsa pecunia crevit" (the love of money grows in proportion to the amount of money itself).

122 Matt 7:3.

123 Juvenal, *Sat.*, 8.140–41, ed. Clausen, 110 (cf. Walther, *Proverbia*, no. 19798).

Emblematic of this is the fact that when Moses the law-giver discussed unclean animals, by which he was referring to the uncleanness of human beings, he was more careful in dealing with the uncleanness of birds. For he counted more birds unclean than fish or mammals, as is clear from Deuteronomy 14.[124] The uncleanness of birds stands for the uncleanness of clerics; that of mammals and fish, the uncleanness of the laity.

Likewise, for other reasons, too, one can assert that the same sin committed by a priest or a cleric is more serious than if it were committed by a member of the laity. For just as a good priest has greater status than a good member of the laity, so, too, he has lesser status if both of them are evil. For if a priest is evil, he cannot exercise the office to which he is bound without sinning, which, however, a member of the laity can do even if he is evil. For this reason, just as a good priest is called a good angel in Malachi 2: "The lips of the priest guard knowledge, and people seek to know the law from his mouth, since he is an angel of the Lord of hosts,"[125] so, too, an evil priest can be called the evil angel. For just as the devil sins in every one of his deeds, even though in itself it may be a good deed – as when he believes God or when he comes out of a possessed body at God's command, since he does all of this with a rebellious murmur and an evil heart –, so, too, an evil priest sins even by carrying out a good deed, as namely by celebrating mass and administering the church's sacraments. And just as it is more dangerous if someone sleeps on a horse than if he sleeps on the ground, because of the danger of falling, so does a prelate sin more seriously and dangerously than someone committed to his care. In an expression of this fact, one reads that the priest Eli fell off his seat and died.[126] And again since those under a prelate's care depend on him as if they were hanging from his neck, if the prelate falls, then as a necessary consequence those in his charge will also topple when he does. For this reason in Isaiah 3 it is said to the prelate: "Be our ruler, and you will be in charge of these ruins."[127]

For we witness among those who go to sea that if one sailor stops rowing, this results in a minor inconvenience to all the rowers, but if the captain fails to steer, this causes the boat to suffer shipwreck and to go under. In the same way, the sin of a prelate who has the duty to guide others is more injurious than that of a person without ecclesiastical office, Ecclesiasticus 10: "A foolish king will destroy his people."[128] And

124 Cf. Deut 14:1–21.

125 Mal 2:7.

126 Cf. 1Sam 4:18.

127 Isa 3:6.

128 Sir 10:3.

therefore in Deuteronomy 17 it is said of the king: "He must not lead the people back to Egypt when the number of his cavalry has increased."[129] Everyone who holds a high position should pay special attention lest the populace committed to him be corrupted by his example, and lest he who should lead the people up to the celestial Jerusalem lead them instead to darkness and into error. Indeed, Egypt is understood to mean "dark places" or "sorrow."[130] If indeed a prelate is responsible for everything, he is also seen to be the author of everything: Because he can correct the evils done by those in his care, he is rightfully a participant in the evils that he was unwilling to amend. For this reason, prelates can and should be very fearful when they hear what the wrathful Lord said to Moses after Israel provoked the Lord by fornicating with Midianite women: "Take all the leaders of the people and hang them from the gallows in the sun, so that my wrath may be turned away from Israel."[131] In fact, one reads that the people sinned and their leaders' fornication is nowhere mentioned explicitly, but nevertheless the leaders are ordered to be led to the gallows, and in their punishment the Lord is reconciled with the people who transgressed because the malfeasance of those in the care of prelates most often comes about from their prelates' negligence. Thus, it is clear that the blemish of sin is more harmful when it stains the eye of the church's* body, i.e., a prelate, than when it soils another of its members, i.e., a person without ecclesiastical office.

And therefore the Creator of nature wisely saw to it that there is no blemish in the sun such as we see in the moon. For according to Bernard Silvestris*:

> The sun is the world's eye since it surpasses all of the
> Stars and lays claim to the heavens as its own.[132]

For the sun is raised up over every other thing, so that it shines on all things with its rays. In just this way a prelate, surpassing everyone like another sun, should pay attention to the merits of each individual. Like the sun he should dry up the mud, i.e., whatever has become moist through lust, and melt what has been frozen through avarice and pitilessness and impiety and arrogance, by the example of his virtuous deeds, as if by his rays. For rarely or never do Holy Scriptures mention a dark-

129 Deut 17:16.

130 Cf. Jerome, *Lib. interp. Hebr. nom.*, Act., ed. de Lagarde, CCSL 72:143. Note, however, that the meaning "sorrow" is not found in Jerome. See above, chapt. 6.1, p. 18.

131 Num 25:4.

132 Bernard Silvestris, *Cosmogr.*, 2.14.41–42, ed. Dronke, 151.

ening of the sun without some memorable lesson and without the darkness of the moon and sometimes that of the stars immediately following that passage. For this reason it is said in Isaiah 13: "The sun has been darkened at its rising, and the moon will not shine with its light."[133] And Joel 2: "The sun and the moon have been darkened, and the stars have withdrawn their light."[134] And again Joel 3*: "The sun will be turned to darkness and the moon to blood."[135] In this we are given to understand spiritually that when a prelate sins, all of those under his care also sin, like someone who has been provoked to sin. For it is said in Ecclesiasticus 10: "Like a people's judge, so also their minister; and like a city's king, so also its inhabitants,"[136] and so on.

133 Isa 13:10.
134 Joel 2:10.
135 Joel 2:31.
136 Sir 10:2.

CHAPTER THIRTEEN

On the Four Things Spiritual Eyes Should Contemplate Continually

A person should turn his attention most often to these four things, namely the filthiness of his own iniquity, the hardship of another person's misfortune, the bitterness of everlasting punishment, and the recompense of a blessed life, so that in this way he is fitted with eyes for every direction like the living creatures John says he observed in Revelation 4: And there were "living creatures, full of eyes front and back,"[1] and so on. For with the right eye a person should turn his attention to rewards; with the left, to punishments; with the one in back, to the misfortune of his fellow human beings and the unfortunate – as it were, the world's rejects; with the one in front, to his own misdeed, according to the Psalm: My sin is always in front of me.[2] As an image of this matter, Moses is told in Deuteronomy 3: "Go up to the summit of Mount Pisgah, and scan with your eyes to the west and to the north, to the south and to the east, and look."[3] By the east [is meant] the eternal rewards that will, as it were, forever rise anew; by the west [he means] the everlasting punishment into which all temporal happiness sinks and is extinguished; by the south [is meant] the heated passion of sin; by the north, the misfortune of one's fellow human being. Thus, each and every person ought to be fitted with multiple eyes, so that he not only looks at present circumstances, but diligently considers the future. For this reason Boethius says in *The Consolation of Philosophy*: "It is not sufficient to contemplate what is in front of your eyes, but prudence ponders how things will end."[4] This forethought or foresight is appropriate especially for a prince or a prelate. For this reason Aristotle said to Alexander in *On the Secrets of Secrets*: "You should be prudent in foreseeing those things that can happen, for when they come about, everyone will know that your eyes see things from a long way off, and because of this they will praise your wisdom and they will beware of offending your sovereignty."[5]

1 Rev 4:6.
2 Cf. Ps 31:5; cf. Ps 49:8.
3 Deut 3:27.
4 Boethius, *Phil. cons.*, 2.pr.1.15, ed. Bieler, CCSL 94:18.
5 Ps.-Aristotle, *Secretum secretorum*, 1.17, ed. Steele, 55.

On paying attention to one's own sin.*

First among the four things enumerated above to which someone should pay attention is his own misdeed. For this reason the sinner is told in Jeremiah 3: "Lift up your eyes to the steep heights and see where you are now* ruined!"[6] For indeed if a traveler raises his eyes and sees a befouled, dangerous, and twisted path, he will turn aside from it, retracing his steps. The path of sinners is befouled because of the filth of lust, dangerous because of the pitfalls of avarice, and twisted because of the distortion of pride, and therefore it is playing it safe to turn back and leave such a path. And for this reason somewhat before this passage the prophet had said first: "Know and see that it is evil and bitter for you to have forsaken the Lord your God!"[7] For a person ought to know by recollecting things in the past, and to see by examining things in the present, that it is evil for him (because he has robbed himself of what God gave freely) and bitter (because he has wounded his conscience) to have forsaken God, who created him when he did not exist, who redeemed him when he died.

Thus, a person should inspect his own misdeed, but do so in order to find it detestable, not pleasant. On occasion, the mere recollection of sin proves lethal. In a sign of this, Lot's wife perished, since she looked back at Sodom which she had left behind, Genesis 19.[8] Therefore, Augustine says: "Oh humankind, if you were to see yourself, you would displease yourself and please me. But now, you who do not see yourself, you please yourself and displease me. The time will come when you will displease both yourself and me equally: me, since you will be judged; you, since you will burn."[9] For sinners do not see the danger of their situation now, nor is that surprising, since they are blind, 2 Kings 6: "Strike this people with blindness."[10] And there follows: "'Lord, open the eyes of these people so that they may see.' And the Lord opened their eyes and they saw that they were in the midst of Samaria,"[11] among enemies. Likewise, if a sinner would open his eyes, he would see that he is in the midst of enemies, i.e., of demons. For the devil puts much effort into first making

6 Jer 3:2. *Pr1* follows the Vulgate text in reading *ubi non prostrata sis* (where you are not ruined), but *P7P8* demonstrate a change in the wording that fits the present context. See below, chapt. 14.3, p. 187.

7 Jer 2:19.

8 Cf. Gen 19:26.

9 Manegold de Lautenbach (?) (Ps.-Bede the Venerable), *Exegesis de Psalmorum libro*, In Ps. 49, PL 93:746C; cf. Augustine of Hippo, *Enarr. in Ps., In Ps.* 49, 28, ed. Dekkers and Fraipont, CCSL 38:596.

10 2Kgs 6:18.

11 2Kgs 6:20.

someone blind so he cannot see his own sin, just as a thief who has broken into a house will normally first put out the light. For this reason one reads that Antiochus entered the temple and carried off the "candelabrum of light," 1 Maccabees 1.[12] For just as Samson, who was deceived by Delilah and taken captive by the Philistines, was first stripped of his hair, after that blinded, and third bound with ropes,[13] in the same way, the sinner, who is deceived by his flesh and taken captive by the devil, is first robbed of his virtues; second, made blind in his interior eyes; and third, tied up by the ropes of sins. And just as a bear, made blind by the glow of a basin-shaped lamp,[14] sports before the eyes of its master, so a sinner, made blind by what he thinks of as pleasures, indulges wantonly in sin before the eyes of demons. For this reason one reads in Judges 16: After the Philistines had blinded Samson, "they ordered that Samson should be called and should make sport before them."[15] And just as the king of Babylon slew the sons of Zedekiah before his eyes and tore out Zedekiah's eyes,[16] so also the king of disorder, the devil, eradicates good deeds through sin and makes a sinner spiritually blind, since after a person ceases to perform good deeds, the light of grace is removed from him and he is blinded internally. Now, sinners are blinded to the extent that they have lost the power of discernment, and they are signified by the people of Sodom, who, as it is said in Genesis 19, having been stricken with blindness, were unable to find Lot's door.[17] For Lot is interpreted to mean "turning aside,"[18] and they do not know how to turn aside from sin.

*On paying attention to the misery of one's fellow human being**.

Second, the misfortune of one's fellow human beings must be considered, Ecclesiasticus 4: "Do not turn your eyes away from a pauper."[19] And again: "Do not avert your eyes from the needy."[20] The author repeats the same idea in order to warn us all the more to feel compassion. For there are few who look on paupers with sympathy, but there are many who avert the eyes of mercifulness from them and cannot see their wretched-

12 1Mac 1:23.
13 Cf. Judg 16:19–21.
14 Cf. above, chapt. 8.5, p. 92.
15 Judg 16:25.
16 Cf. Jer 39:6–7.
17 Cf. Gen 19:11.
18 Jerome, *Lib. interp. Hebr. nom.*, Luc., ed. de Lagarde, CCSL 72:140.
19 Sir 4:1.
20 Sir 4:5.

ness. For an eye which is in the light does not see someone who is in darkness. So, too, the rich man who is in the light of affluence does not see the pauper who is in the gloom of misfortune. And for this reason as long as the chief butler was in the darkness of prison, he acknowledged the mistreated Joseph and was his companion, but after the butler had gotten out into the brightness of prosperity, he did not know Joseph any longer, but rather, "once a time of good fortune followed, the chief butler forgot the interpreter of his dream," Genesis 40.[21] I understand that when a certain lady had seen paupers in a field collecting plants to eat during a time of famine, she said, "Surely they could eat peas with lard, couldn't they?" A man* standing nearby responded, "No doubt they could, if they had them." Indeed, since this woman's eye was located in the light of affluence, she did not see the indigence of the poor persisting in darkness. For:

> A full belly does not know what an empty one would seize on.[22]

We see that if someone who has been in the light for a long time approaches a person suffering in darkness, he does not acknowledge him at first until after he has been with him for a while. And in the same way, someone who spends his life in the light of affluence does not look with an eye of compassion on people who are located in the darkness of misfortune, nor does he recognize them, nor does he have pity on them, until he joins with them in sympathy so that he suffers with them in his mind in the same way that they suffer in their bodies. These matters are represented in Exodus 3, where one reads that as Moses was about to go to see a great vision, the Lord said, "Remove the shoes from your feet,"[23] as if he were saying, "You will not be able to see this vision unless you first put your shoes to one side." After he put them to one side, he approached to see the vision – why the bush was burning and was not consumed by the flames. The burning bush that is not consumed by the flames is a person poor and sick in poverty and misfortune, not one crushed through lack of patience but who remains unmoved because of his patience, something which wealthy people wearing their shoes consider a miracle and a great vision. To see this bush with a compassionate eye is not possible for a wealthy person, crammed full of riches and pleasures and not consumed by misfortunes, unless he first removes the shoes that protect his feet from impediments, i.e., the provisions for his flesh, which hinder the mind's good-will so that it does not go on to compassionate deeds. And

21 Gen 40:23.

22 *Sententia communis* (Walther, *Proverbia*, nos. 16561 and 16562).

23 Exod 3:5.

for this reason the chief butler, who was mentioned earlier, was as it were barefoot in prison, i.e., he was separated from pleasures: he clearly saw the bush that was on fire and was not consumed by the flames, i.e., Joseph, who suffered but was not led astray. But after the chief butler put on shoes of prosperity and pleasures again, he no longer recognized Joseph anymore.[24]

Do you want to see someone who is generous to the poor and never turns his eye from their needs? Observe the blessed Job and hear what a just man and one who "feared God and turned away from evil"[25] says. He does not say, "Actors and mimes converged on me, sycophants and zealous seekers of useless fictions." He does not say, "I have kept bears; I have trained monkeys and young dogs." Rather, in Job 31 he says, "Let my shoulder fall out of its socket, and let my arm be crushed with all its bones"[26] "if I have denied the poor what they wanted, if I have caused the eyes of the widow to wait with apprehension, if I have eaten my morsels alone and the orphan has not eaten of them."[27] And a bit earlier he says, "I was an eye to the blind and a foot to the lame. I will be a father for the poor" and not a robber, Job 29.[28] And so that you will have all the more abundantly a heart of compassion, observe also another person who in a resemblance of the blessed Job did not cause the eyes of the widow to wait with apprehension. In the *Deeds of the Romans* it is recounted that when Trajan had already mounted his horse and was about to march out to war with his army, a certain very poor widow caught his foot and called out to the emperor saying, "Lord, give me justice for the one who killed my son." He said to her, "When I return, I'll show full justice." And she said, "What if you don't return?" And he said, "My successor will give you satisfaction." She said to him, "It's you who are indebted to me; you will receive according to what you deserve, and it's deceit not to give back what's owed. Your successor will be rewarded for a gracious judgment, but another's justice will not release you." Moved by these words, the emperor got off his horse and made his entire army wait: He sat at the head of the tribunal, diligently looked into the widow's case, and alleviated her grief with commendable satisfaction.[29] Therefore – and this has

24 Cf. Gen 40:1–23.

25 Job 1:1.

26 Job 31:22.

27 Job 31:16–17. The text from the beginning of the paragraph through this citation from Job has been taken, partially verbatim, from John of Salisbury, *Policr.*, 8.13, ed. Webb, vol. 2, 322–23.

28 Job 29:15–16.

29 Cf. John of Salisbury, *Policr.*, 5.8, ed. Webb, vol. 1, 317–18 (cf. Tubach, *Index exemplorum*, no. 4989).

been spoken of as a miracle – when God observed him with the eye of his compassion, he recalled him from hell at the entreaties of the blessed Gregory.[30] "Blessed are the merciful, for they will obtain mercy."[31] This man could truly say with the blessed Job: "If I have caused the eyes of the widow to wait with apprehension"[32] (Job 31), "I have also comforted the widow's heart"[33] (Job 29).

Then follows paying attention to punishment in hell.*

Third, punishment in hell must be considered. After the prophet Jeremiah had seen "a boiling pot and it was facing away from the direction of the north," and after he had heard the Lord saying, "From the north all evil will spread out," Jeremiah 1,[34] he then urged others to gaze attentively at the same sight, Jeremiah 13: "Lift up your eyes and see, you who are coming from the north,"[35] and so on. By the north, the most inclement region and the one furthest from the sun, we are fittingly given to understand punishment in hell, where one will find a bitter punishment and the separation from the sun of justice because divine vision is missing there. This punishment was well observed by Jerome who said: "When I consider Job sitting in the dung-pit,[36] John suffering hunger in the wilderness,[37] Paul slain by the sword,[38] Peter hanged on the gallows,[39] then I think about what he will do in the future to the damned if he afflicts people he loves in such a way in the present world."[40]

This sight is useful, both for avoiding a misdeed and for reflecting on the mildness of the present punishment. For this reason Isidore says in book 1 of the *Soliloquies*: "Lay out for yourself the fire of future torments as a way to counteract the ardent desires of the flesh: Recalling to mind

30 On the legend of Trajan and Pope Gregory the Great, see Tubach, *Index exemplorum*, no. 2368.

31 Matt 5:7.

32 Job 31:16.

33 Job 29:13.

34 Jer 1:13–14.

35 Jer 13:20.

36 Cf. Job 2:8.

37 Cf. Luke 1:80.

38 Cf. Eusebius of Caesarea, *Chron.*, trans. Jerome, anno 2084 (from Abraham), ed. Helm, 267 (in Jerome).

39 Cf. ibid.

40 Cf. Thomas of Ireland, *Manipulus florum*, Tribulacio AC, ed. Chris Nighman, available at http://info.wlu.ca/~wwwhist/faculty/cnighman/index.html (accessed December 10, 2010), where the quotation is attributed to Gregory.

hell's burning heat drives out the burning passion of lust."[41] We see that when someone burns his finger, he brings it close to the fire so that the lesser heat can be drawn out by the greater one. So, too, if a sinner inflamed by the fire of sin brings himself close to the fire of hell through self-reflection, the lesser heat can be drawn out by the greater one, Ecclesiasticus 7: "Remember the end awaiting you," i.e., the torments of hell, "and you will never sin."[42] This contemplation also makes someone think his present punishment is very mild. For this reason Bernard says: "You fear the vigils and fasts and the manual labor, but these are mild to anyone considering eternal flames."[43] Bede relates in the *Deeds of the English* that after a certain proprietor of an estate had died in England at the time of Emperor Constantine the Younger, the proprietor's soul was led before the dais of the everlasting judge where at last through the intercession of God's mother he was allowed, first, to witness the pains of hell and, afterwards, to return to his body. Thus, he who had died came to life again, and after he had related the wonders he had seen and had disbursed his property to the poor and had paid his respects to his relatives as well as* his servants, he departed for a certain forest where he performed such penance that he demonstrated in word and at the same time in deed how great the torments were that he had witnessed. Even at the height of winter he used to immerse himself naked in a river, and when his friends and acquaintances rebuked him for this, he would respond, "Leave me be, for I have seen more severe things."[44]

The contemplation of punishment in hell causes the mind's eye to open, for according to Gregory "punishment opens the eyes which a misdeed closes."[45] And it not only does this when it is experienced (as is clear in the case of the rich feaster who, "after lifting up his eyes when he was in the midst of torments,"[46] saw Lazarus, from whom he had turned his eyes earlier when he was in the midst of wealth), but it also does so when it is diligently examined with the eye of contemplation. For just as a salve clears up the eyes, so does the contemplation of the torments of hell make things clear to the heart and mind. And for this reason Job's

41 Isidore of Seville, *Synonym.*, 1.46, ed. Elfassi, CCSL 111B:37–38 (PL 83:837D–38A).

42 Sir 7:40.

43 Bernard of Clairvaux, *Ep.*, 1.12, ed. Leclercq and Rochais, in *Opera*, vol. 7, 10.

44 Cf. Bede the Venerable, *Historia ecclesiastica gentis Anglorum*, 5.12, ed. and trans. Bertram Colgrave and Roger A.B. Mynors, in *Bede's Ecclesiastical History of the English People* (Oxford, 1969), 488–98 (cf. Tubach, *Index exemplorum*, nos. 1818 and 2512).

45 Gregory the Great, *Mor. in Iob*, 25.5.6, ed. Adriaen, CCSL 143B:1233.

46 Luke 16:23.

eye was so sharp-sighted that he could say to the Lord, "Now my eye sees you," Job 42,[47] as if he had anointed it with a salve so that he could see. For he says in Job 17: "My eye tarries in bitterness."[48] A salve is something sharply bitter* for the eye. Thus, that person's eye tarries in bitterness who diligently considers the bitterness of punishment in hell and through the contemplation of this bitterness he recovers the vision he had lost. And so as a symbol of this, one reads that Tobit recovered his sight by an anointment of gall, Tobit 11.[49] And therefore sinners, who do not think about the torment of hell, are like condemned prisoners who are led blindfolded to the gallows. For he is blind who does not look at his own sin so that it displeases him, nor at the reward of heaven so that it pleases him – and likewise he is blind who does not see eternal punishment so that it deters him. And it is surprising that people do not give thought to the sufferings in hell, or, if they do, that these sufferings do not terrify them, since even "the demons believe, and they quake with fear," as it is said in James 2.[50] One reads in the *Lives of the Fathers* that when the spirit of fornication was sorely tempting a particular saintly father, he prayed to the Lord that the tempting spirit should appear to him visibly. Then a demon appeared to him with a human face. When he saw him, the holy man said, "What profit do you get from tempting me? For this is the greatest folly. Surely you know that when you bring someone down in sin, you increase your own sin and, as a result, your punishment. Nevertheless your zeal is aimed completely at bringing people down in sin, and thus, as a result, at the accumulation of your punishment." In response, the devil said, "You should know what you say is true, but I'll tell you the reason for my tempting. I know that the more people I cause to sin, the longer I'll delay the day of judgment, and I dread that day more than anything, on which I expect to hear the hard sentence: 'Go, cursed ones, to the eternal fire which has been prepared for the devil and all of his angels.'[51] And since I'm able to delay for at least a while the calamity of that sentence, for that reason I strive so much to tempt people."[52]

On paying attention to life in heaven.*

Fourth, one must pay attention to the reward of life in heaven. For this reason in Hebrews 11 it is said in reference to Moses: "For he looked to

47 Job 42:5.
48 Job 17:2.
49 Cf. Tob 11:8–13.
50 Jas 2:19.
51 Matt 25:41.
52 Cf. Tubach, *Index exemplorum*, no. 2573.

the reward."[53] And therefore it is said in Isaiah 33: "Your eyes will see Jerusalem, an opulent dwelling-place."[54] This is the city that Bernard describes in the *Meditations*: "Oh celestial city, secure habitation, a country containing all things that are pleasing, a populace without complaint, its inhabitants at peace, its people wanting nothing. How 'glorious are the things spoken of you, city of God!'[55] 'Dwelling in you is, as it were, a matter of all rejoicing.'[56] There is:

> Peace, piety, goodness, light, virtue, magnificence, integrity,
> Joy, the sweetness of rejoicing, life everlasting,
> Glory, praise, calm, love, and the harmony of peace.[57]

Thus, what madness compels us to drink the wormwood of vices, to pursue the shipwreck of this world, to bear the lordship of an ungodly tyranny, and not rather to fly to the company of angels, the splendor of the saints, the esteem of a powerful king, the beauty of celestial glory?"[58] For this reason the eye of our heart should be fixed on heavenly matters through meditation and longing, something the upright posture of the human body also teaches. And that is what the poet says:

> And while the other animals, bent over, look at the earth,
> He gave humankind a lofty countenance, and commanded him to
> View the heavens and to lift his upright gaze to the stars.[59]

or this reason, too, when Empedocles was asked why he had been born, he responded: "So that I might contemplate the heavens and the stars." Bernard Silvestris also recollects his saying when he writes:

> To one inquiring why he was alive Empedocles said, "To gaze
> At the stars; take away the heavens and I'd be nothing."[60]

One reads in the *Lives of the Fathers* that when the blessed Arsenius was dying, he was asked by the brothers to leave an edifying sentence to

53 Heb 11:26.
54 Isa 33:20.
55 Ps 86:3.
56 Ps 86:7.
57 *Sententia communis* (Walther, *Proverbia*, no. 21054).
58 Ps.-Bernard of Clairvaux, *Medit. piissimae*, 4.11–12, PL 184:492C–93A (cf. Vincent of Beauvais, *Spec. hist.*, 28.21, in *Speculum quadruplex*, vol. 4, 1149a).
59 Ovid, *Met.*, 1.84–86, ed. Anderson, 4 (cf. Walther, *Proverbia*, no. 22635).
60 Bernard Silvestris, *Cosmogr.*, 2.14.45–46, ed. Dronke, 151.

remember him by before he went to the Lord. He said, "There, where." And once he said this, he gave up the ghost. After he passed away, the brothers were eager to understand the meaning of his words and sent for an explanation to a certain other man advanced in years who wrote back to them: "This, I believe, is the meaning of his words: Our hearts are bound fast there where our true joys lie."[61] Jerome exhorts the virgin Eustochium to this thorough examination of celestial matters by saying in one of his letters: "Come out for a little while from your prison, and picture before your eyes the reward for your present labor, which 'the eye has not seen nor the ears heard.'[62] What will that day be like, when Mary, the mother of the Lord, will meet you accompanied by choirs of virgins, when, with the Red Sea behind her and pharaoh drowned with his army, holding a timbrel* she will sing before those who respond, 'Let us sing to the Lord, for he is gloriously honored,'[63] and so on? At that time, too, your bridegroom himself will meet you and say, 'Arise, hurry, my love,'[64] and so on. At that time, the angels will also marvel and say, 'Who is she who goes forth like the dawn?'[65] Daughters will see you, and 'they will implore you,'[66] and so on. Sarah will come to meet you with the wives, Anna with the widows. Children will also call out, 'Hosanna in the highest.'[67] As often as the empty flattery of the world delights you, as often as you see something to glory in, ascend in your mind in this way to paradise."[68]

61 Cf. Tubach, *Index exemplorum*, no. 358.

62 1Cor 2:9.

63 Exod 15:1.

64 Cant 2:10.

65 Cant 6:9.

66 Cf. Ps 44:13.

67 Mark 11:10.

68 Jerome, *Ep.*, 22.41.1–5, ed. Hilberg, CSEL 54:209–10.

CHAPTER FOURTEEN

On Three Visible Things That Delight the Corporeal Eye

There are three things that the corporeal eye looks at with delight, namely the translucence of water, the beauty of flowers and plants, and the radiance of a mirror. For this reason the verse reads:

> Delightful to the eye: clear water, verdure, a mirror.[1]

By water I understand for the present purpose the glorious Virgin; by flowers and plants, the company of saints; by the mirror, our Lord Jesus Christ. For indeed the Blessed Virgin is called "a garden fountain, a well of living water," Song of Songs 4.[2] For just as water is the common possession of all, according to the verse:

> Water is for the benefit of all,[3]

so also the Virgin is blessed for the just and the sinners. For this reason the blessed Bernard said: "Mary was made all things for all people so that everyone would receive something from her fullness: prisoners, their release; the diseased, their cure,"[4] and so on. And Anselm says in the *Meditations*: "Oh Lady, the joyful assembly of the just comes to you and praises you, the frightened crowd of the guilty flees to you. If only my heart could sleep in your love, my soul melt, my flesh disappear!"[5] For next after God, the most important thing is to see you and contemplate

1 *Flos medicinae Scholae Salerni*, 9.17.1959, ed. Salvatore de Renzi, in *Collectio Salernitana: Ossia documenti inediti, e trattati di medicina appartenenti alla scuola medica Salernitana, raccolti ed illustrati da G.E.T. Henschel, C. Daremberg, e S. de Renzi* (Naples, 1852; reprint 2001), vol. 1, 509.

2 Cant 4:15.

3 Ovid, *Met.*, 6.349, ed. Anderson, 135.

4 Bernard of Clairvaux, *Sermo in Dominica infra octavam assumptionis*, 2, ed. Jean Leclercq and Henri M. Rochais, in *Opera* (Rome, 1968), vol. 5, 263.

5 Anselm of Canterbury, *Orationes*, 6 and 7, ed. Franciscus S. Schmitt, in *Opera omnia* (Edinburgh, 1946), vol. 3, 15, 24.

you, to delight in your praises. For this reason Mary reveals herself to all in the likeness of water, saying in Ecclesiasticus 24: "Cross over to me, all you who desire me, and be filled with my fruits,"[6] as if she were saying what is written in Isaiah 55: "All those who thirst, come to the waters,"[7] and so on. The eyes of the blessed soul are all fixed in delight on the sight of this water, according to the passage in Song of Songs 5: "His eyes are like doves beside brooks of water that linger near rivers flowing at full strength."[8] Nor is it surprising if the sight of her brings delight, for "the eye will gaze in wonder at the beauty of her radiance," as it is said in Ecclesiasticus 43.[9] She is Queen Esther, about whom it is said: "She was exceptionally becoming and of unparalleled beauty, finding favor in the eyes of all,"[10] Esther 2.

There was a certain cleric who was very devoted to the most Blessed Virgin and who gave careful attention to what is written in Song of Songs 4: "How beautiful you are, my love, sweet and elegant."[11] And the verse in chapter 7: "How beautiful you are, and elegant in delights, dearest."[12] Hearing that all these passages are understood and explained as references to the Blessed Virgin, he asked her to consider him worthy enough to appear to him. He thought he would be blessed if just once he were able to look at someone so beautiful. After a long period of time, an angel came to him appointed on behalf of the Blessed Virgin and saluting the cleric said to him, "The Blessed Virgin has heard your prayer and sends word that she'll come to you on a particular day and hour and you'll see her. But you must know that after your eyes have seen her beauty and splendor that exceeds all others, it's extraordinarily improper and shameful for them to see anything worldly and transitory." The cleric said, "I don't care if I become blind from then on after I have beheld her once." But after the angel had left, out of weakness the cleric became very concerned, and he thought about what he would do after he had lost his eyesight. "For the rest of my life," he said, "I won't be able to earn my bread by writing; rather, I'll wind up even more miserable and beggarly, I'll be without all comfort, and I'll be able to say with Tobit, 'What kind of joy do I have, since I sit in darkness and do not see the light of heaven?'"[13] At

6 Sir 24:26.
7 Isa 55:1.
8 Cant 5:12.
9 Sir 43:20.
10 Esth 2:15.
11 Cant 4:1 and Cant 6:3.
12 Cant 7:6.
13 Tob 5:12.

length he considered and said, "When the Blessed Virgin comes, I'll open one eye to see her and shut the other, and in this way I'll lose only a single eye, and since I'll have one eye left over, that'll be enough for me." But when the blessed and glorious Virgin appeared at the designated hour, he beheld such beauty and splendor that it can be neither expressed nor contemplated, and when he removed the hand he had placed over* one eye (so as not to see her with it) in order to open that eye and thus behold her more completely, the glorious Virgin disappeared and he did not see her any more. Then he was deprived of one eye (the one, that is to say, with which he had seen her) and the other one was whole (with which he had not beheld her), but he rebuked himself bitterly and was very dejected, saying through his tears, "Alas, I'm miserable, why did I shut one eye? Why didn't I open both? Oh, if only I'd been all eyes so that I could've seen her more completely!" And he begged the Blessed Virgin for a long time to reveal herself to him once again so that he could look at her entirely, at least with his remaining eye, wishing to be deprived of both eyes in order to gaze upon her once more. Then the angel was sent to him again and said, "My Lady, whose messenger I am, sends word to you: 'Friend, why do you seek more? Do you want to be deprived of your remaining eye by seeing me* once again?'" He said to the angel, "By all means, sir, even if I had a thousand eyes, I would choose to lose them for all time, so that I could see her completely." The angel said to him, "And she who is most merciful sends word to you that you'll see her again, and you won't lose your remaining eye at all, but rather you'll actually recover the lost one." And this is what happened not much later, for once again the gentlest Virgin appeared and revealed herself to him to be seen and contemplated and she restored his lost eye to him.[14] From this we can judge how great the glorious Virgin's beauty is and how delightful it is to see her with one's eyes. For this reason Anselm also says in the *Meditations*: "Mary is beautiful to look at, lovely to embrace, charming to delight in."[15]

*The second thing that delights vision signifies that one should carefully consider the verdant examples of the saints**.

The second thing that delights the eyes is the beauty of flowers and the loveliness of plants. For this reason it is said in Ecclesiasticus 40: "Your eye will desire grace and beauty, and more than this, green fields growing."[16] Eccle-

14 On this narrative, cf. Tubach, *Index exemplorum*, no. 4365.

15 Anselm of Canterbury, *Orat.*, 7, ed. Schmitt, vol. 3, 21.

16 Sir 40:22.

siasticus 39: "Bloom, flowers, and grow verdant in grace!"[17] For since the saints blossomed internally by their holy vows and blossomed externally by their exemplary deeds, therefore they are pleasing for the eye to see.

Indeed, the eye of the righteous man is delighted when it perceives the roses of martyrs, the violets of confessors, and the lilies of virgins. For according to Pliny, the rose blossom maintains a position of pre-eminence among the other flowers, and therefore it is customary that the most important part of human beings, namely the head, is crowned with rose blossoms.[18] For the rose blossom is beautiful to the sight and fragrant to the sense of smell; indeed, by its beauty it restores the sight and by its fragrance it attracts the sense of smell, and the pre-eminence of the martyrs is appropriately designated by it. And for this reason a flower-basket filled with blood-red roses adorned the coffin of the first martyr Stephan, as the account of the discovery of his relics has it.[19] And in Ecclesiasticus 50 this is said concerning Simon, whose name is interpreted to mean "obedient,"[20] and who stands for any martyr who was obedient to God the father to the point of death: "Like a rainbow he gleamed among the clouds of glory, and like rose blossoms on spring days."[21] For a rainbow gleams in the midst of the clouds, and the holy martyrs have a reddish glow in the midst of their torture, and they are said to be among the clouds of glory since while they were being tormented, their thoughts were on heavenly glory. And they are said to smell sweetly like blossoms on spring days because of their re-enactment of the Lord's passion which took place during the springtime.

When the violet blossom reaches maturity, it bends its head and lets it hang down, and it compensates for the slightness of the material it is made of with the grandeur of its virtue and fragrance, and it stands for the holy confessors who were exceedingly humble. For this reason one of them says: "Whoever gathers together the rest of the virtues without humility, it is as if he were carrying dust in the wind."[22] And another

17 Sir 39:19.

18 Cf. Pliny the Younger, *Nat. hist.*, 21.4(10).14–21, ed. von Jan and Mayhoff, vol. 3, 385–87.

19 Cf. Stephanus Leodiensis, "L'office de l'Invention de S. Etienne," ed. Ritva Jonsson, in *Historia. Études sur la genèse des offices versifiés*, Acta Universitatis Stockholmiensis. Studia Latina Stockholmiensia, 15 (Stockholm, 1968), 214.

20 Jerome, *Lib. interp. Hebr. nom.*, Act., ed. de Lagarde, CCSL 72:148. The etymology of Hebrew *shimon* is doubtful; cf. Gen 29:33 where the name is derived from Heb. *shama* (he heard), a linguistically correct derivation for the first letters of the name.

21 Sir 50:8.

22 Gregory the Great, *Expositio in septem psalmos poenitentiales*, 3.3, PL 79:569C.

one says: "From my youth there is nothing I have tried to avoid as much as a puffed up heart and an unbowed neck, things that provoke God's hatred against them. For I know the Lord my master said in the humbleness of his flesh, 'Learn from me, since I am gentle and humble in heart.'"[23]

But according to Pliny nothing is more agreeable in color, more sweet-smelling in its odor, and more powerful in its virtue than the lily blossom.[24] The lily blossom has this property: As long as it remains undefiled and untouched, it gives off a sweet smell, but once it is broken off and chafed by the hands, it reeks horribly.[25] And it appropriately stands for virginity: Nothing is more agreeable than virginity, nothing more fragrant, as long as it remains unimpaired, but it reeks to God and humanity if it has been violated through unchasteness. Thus, the saints blossomed during their earthly pilgrimage by the purity of their way of life*, and they blossom now in the heavenly homeland by attaining a state of blessedness for the soul.[26] For this reason the Psalm says: "They were planted in your house,"[27] Lord, and so on. Finally, they will also blossom in their body through the gift of incorruptibility. For this reason the Psalm says: "They will blossom from city" to city "like hay from the earth,"[28] i.e., in the likeness of blades of grass. For just as grass dies during the winter and is reborn at the time of renewal, so also the body of a human being is reduced to dust in this present life and in the future one will rise again in incorruptibility.

Clearly, that person who is not delighted by the sight of flowers of such great beauty appears to be spiritually blind. And just as the appearance of flowers and plants invigorates and strengthens corporeal sight, so also the life and the glory of saints increases the spiritual strength of those who regard them attentively. For this reason Gregory says: "If we reflect on the deeds of those who came before us, the things which we endure will not be burdensome."[29] Thus, let us concentrate on the roses of the martyrs' patience and rid ourselves of laziness and bear the fruits

23 Matt 11:29; Jerome, *Ep.*, 76.1.1–2, ed. Hilberg, CSEL 55:35.

24 Cf. Pliny the Younger, *Nat. hist.*, 21.5(11).22–23, ed. von Jan and Mayhoff, vol. 3, 387–88.

25 Cf. Bartholomeus Anglicus, *De prop. rer.*, 17.91, sig. z2ra (Middle English trans. by Trevisa, ed. Seymour et al., vol. 2, 982).

26 Peter uses the common trope that contrasts the earthly journey *in via* (on the way) with reaching the end of that journey *in patria* (in the heavenly homeland).

27 Ps 91:14.

28 Ps 71:16.

29 Cf. William Peraldus, *De erud. prin.*, 5.35, available at http://www.corpusthomisticum.org/xreo.html (accessed December 10, 2010).

of good deeds. Ecclesiasticus 39 advises us to do this: "Listen to me, Godly fruits, and be fruitful like a rose planted by streams of water."[30] Let us also concentrate on the violets of the humble confessors and reject pride and be pleasing to the Lord through the sweet smell of humility, so that we can say the verse in Song of Songs 1: "While the king was on his couch, my nard gave off its odor,"[31] and so on. Nard is a moderate-sized and aromatic plant and as the *Gloss* says it signifies humility, which is most agreeable to God.[32] Let us also concentrate on the lilies of virgins and cast off uncleanliness and put on sanctity, Ecclesiasticus 39: "Bear flowers, and like the lilies give off your odor!"[33]

Thus, whoever wants to restore the eyes of his mind ought to raise them to the saints and say with the Psalmist: "I have raised my eyes to the mountains from where help will come to me,"[34] i.e., to the saints exalted through the loftiness of a heavenly way of life. In these mountains he will discover such pleasantness that he will despise worldly comforts, in the manner of the wild donkey, about which Job says in chapter 39: "He scorns the masses in the city; he looks around the mountains for pasture, and he searches for every growing plant."[35] For on the mountain of the martyrs he looks around for the pasture of patience; on the mountain of the confessors, for the pasture of humility and obedience; on the mountain of virgins, for the pasture of chastity.

Because of what has already been said, the saints are appropriately symbolized by the green branches partly stripped of bark that Jacob "placed in the water troughs." At the sight of them the sheep conceived speckled offspring, as one reads in Genesis 30.[36] These branches are the saints who were elevated to the heavens, insofar as they were able, although they were on the earth. They are called green since they were always young and fresh in performing good deeds, and when they achieved the greatest perfection, they considered themselves still in their beginning stages. They are said to have the bark partly stripped off since their holiness has not been revealed to us completely, but only partially. And many more things are lying hidden within them internally than are apparent to us externally, Song of Songs 4: "Besides what lies hidden internally."[37] These branches are placed in water troughs since they have

30 Sir 39:17.
31 Cant 1:11.
32 Cf. *Gloss. ord.*, Cant. 1:11, ed. Rusch, vol. 2, 710a in margin (PL 113:1134A).
33 Sir 39:19.
34 Ps 120:1.
35 Job 39:7–8.
36 Cf. Gen 30:37–41.
37 Cant 4:1.

been entrusted to texts about the saints, and a knowledge of the saints flows to us, as it were in troughs of water, by texts of this kind. We ought to have the way of life of the saints before our eyes so that by emulating them we are able to conceive the speckled offspring of good deeds. Now, after the eye of our mind has examined the splendor that the saints possessed on their earthly pilgrimage, this eye should be raised so that it can be delighted more fully by gazing also on the splendor that the saints themselves behold now in their heavenly homeland. About this, Hugh says in book 4 of *The Soul's Cloister*: "The verdure of the meadow that is in the midst of the heavenly cloister invigorates the eyes of its divine residents and always strengthens them with its freshness. For nothing will be aged or fleeting there, but rather without doubt it will remain forever in the freshness of its verdure. The tree of life is located in the middle of the meadow to ensure a pleasurable life. For the tree of life is the wisdom of God the father; its fruit maintains health and preserves life. When a person reaches this tree, he will have nothing further to seek. For iniquity will cease then, love of one's fellow human beings will abound, nor will anything be wanting where each and every person possesses all things by possessing them in himself. Nothing could be lacking when the wisdom of God will look after all things for all people. No one will be lacking anything there within his own station, since each and every person will read in the book of divine wisdom, in which the order of nature has been inscribed."[38]

On the third thing that delights vision. That we should frequently look in a mirror, i.e., Jesus Christ.*

The third delight for the eye is looking in a mirror. In book 2 of *On Mirrors* it is shown through demonstration that the eye is naturally disposed to take delight in a mirror, for it is only in a mirror that it observes itself, insofar as the rays of light are collected together at the endpoint of contact and are reflected in the surface of the mirroring body.[39] And likewise, the contemplative eye, taking delight in the mirror of eternity, never ceases to reflect upon itself, and it continually cleanses blemishes in the conscience by contemplating the purity and cleanness* which it has observed in that mirror. About this mirror it is

38 Hugh of Saint-Victor, *De claust. an.*, 4.34, 35, 36, 42, PL 176:1172D, 1173A, C, D, 1174C, 1176C, 1181B.

39 It is unclear which book on mirrors Peter refers to here; cf. Schleusener-Eichholtz, "Naturwissenschaft," 264n25, 287.

said in Exodus 25: "Look at it, and act according to the pattern which has been shown to you on the mountain."[40] And therefore in Jeremiah 3 it is said to the soul itself: "Lift up your eyes to the steep heights," that is to say, to look in this mirror, "and see where you are now ruined!"[41] For since God and the soul are in the same image, when the soul contemplates God's purity, it carefully considers its own vileness. And therefore, after Job had looked in this mirror, he considered his faults and rebuked himself. And therefore he said, "Now my eye sees you. And for that reason I despise myself and I do penance in dust and ashes."[42] In common parlance it is said that an ugly person does not willingly gaze in a mirror lest he be saddened by perceiving his own ugliness. So, too, many who have persisted in their sins and unclean acts do not willingly look into this mirror, but rather turn away from the sight of it, and in this way they pay no heed to their unclean acts in order to stop doing them. And therefore in Jeremiah 2 the Lord says, "They have turned their back to me and not their face."[43] And after a few verses, this one follows: "How very vile you have become, going back over the same ways!"[44] And why is it surprising if the face that did not look in the mirror to discern its own blemishes has become vile? And therefore people who neither hear God's words willingly nor look at his deeds and achievements with their eye become disgusting in their sins like "pack animals standing in their dung."[45]

And therefore we ought to look in that mirror frequently so that we can discern in ourselves the blemishes of our mind and cleanse them. And therefore it is said in the Psalm: "Look into the face of your messiah!"[46] Blessedness, human and angelic, consists in the unobstructed view of this mirror, John 17: "This is eternal life: that they know you, God alone,"[47] and so on, and 1 Peter 1: "On whom the angels desire to look."[48] Nor is it surprising that that mirror is of the purest quality: "For it is the glory of eternal light, and a mirror without blemish," Wisdom 7.[49] It is

40 Exod 25:40.

41 Jer 3:2. Note that the Vulgate text reads *ubi non prostrata sis* (where you are not ruined). See above, chapt. 13.1, p. 171.

42 Job 42:5–6.

43 Jer 2:27.

44 Jer 2:36.

45 Joel 1:17.

46 Ps 83:10.

47 John 17:3.

48 1Pet 1:12.

49 Wis 7:26.

also something that is able to depict all things. Gregory says: "What is it that they do not see, they who see the one seeing all things?"[50] And therefore the blessed in the homeland who have been made watchmen of his magnitude for this kind of observing praise God and exalt him with praises. For this reason it is said in Isaiah 52: "The voice of your watchmen: They have lifted up their voice, and likewise they will praise, since they will see eye to eye."[51]

50 Alan of Lille, *De fide catholica contra haereticos libri iv*, 1.73, PL 210:374D (cf. Gregory the Great, *Mor. in Iob*, 19.12.20, ed. Adriaen, CCSL 143A:971).

51 Isa 52:8.

CHAPTER FIFTEEN

On the Sevenfold Eye of the Divine Gaze

John says in Revelation: "I looked and, behold, in the midst of the throne there was a lamb standing, as if it had been slain, with seven horns and seven eyes."[1] In the slain lamb we understand Christ, who was most innocent and gentle, and who was sacrificed for us on the altar of the cross. In the seven eyes of the lamb are understood the seven properties of the divine gaze.

That the divine eye looks at everything.*

First, the divine eye is one that can observe all things, the Epistle to the Hebrews: "All things are open and laid bare to his eyes,"[2] and Ecclesiasticus 23*: "His eye sees all things."[3] Peter Damiani* relates that a certain man stole his neighbor's pig. Now, this man was exceptionally hospitable and since "the merciful are blessed because they will obtain mercy,"[4] the Lord Jesus, who never forgets compassionate people, instantly appeared to him in the likeness of a poor man. And since he had long hair, he asked for a barber*. The man immediately stood up with reverence before a pauper, picked up his scissors, and began to give him a haircut. When he did this, he discovered two eyes concealed in the back of the pauper's head. The sight of them made him intensely frightened and he asked what this was supposed to mean. Jesus said to him, "My name is Jesus. I observe everything everywhere. And these are the eyes with which I've seen the pig that you hid in the cave." Then immediately he disappeared, and the man felt remorse and gave back the pig.[5] And thus it is said in

1 Rev 5:6.

2 Heb 4:13.

3 Sir 23:27.

4 Matt 5:7.

5 Cf. Peter Damiani, *Epistolae*, 70, ed. Kurt Reindel, in *Die Briefe des Petrus Damiani*, MGH, Die Briefe der deutschen Kaiserzeit, 4/2 (Munich, 1988), vol. 2, 318–19 (= *Opuscula varia*, 42.5, PL 145:671D–72A) (cf. Tubach, *Index exemplorum*, no. 1018).

Ecclesiasticus 23: "The eyes of the Lord are much brighter than the sun, observing the ways of men in their hiding places."[6]

The divine eye's second [property of] sight, by which it inspires awe, with the example of the sun.*

Second, the gaze of God's eyes is awe-inspiring and because of this it causes people to refrain from sinning. For a thief would not dare commit a theft if he knew he was being watched by a judge, Proverbs 20: "A king who sits on the seat of judgment* puts all evils to flight by a glance,"[7] and Amos 9: "Behold, the eyes of the Lord are upon the sinful kingdom, and I will destroy it from the face of the earth."[8] No one would dare commit fornication in front of other people's eyes; how much more disturbing should it be to perform any misdeed in God's sight! Bernard: "Oh mankind, in every location, in every district, in every corner, show respect for your angel who watches over you, so that in his presence you would not do something that you would not dare do in my sight."[9] This is the angel of great counsel to whom every corner is accessible.[10] And someone has said:

> When you do something evil that would make you blush if I were to see you,
> Why do you not blush all the more about it since God sees you?[11]

Thus, in all transgressions that we commit by thought or word or deed, we ought to fear the gaze of the highest judge, since even if our vices are sometimes not apparent to other people, to his eyes "all things are open and laid bare."[12] And just as we see that where the sun turns its gaze, it dries out the mud, but where it does not gaze, mud remains plentiful, so too when people pay no heed to God's presence, thinking he does not see the disgraceful deeds they commit, they spend their life in the filthiness of sin. For this reason the Psalmist says: "God is not before his eyes; his paths have been defiled for all time."[13]

6 Sir 23:28.
7 Prov 20:8.
8 Amos 9:8.
9 Bernard of Clairvaux, *Serm. in Ps. "Qui habitat,"* 12.6, ed. Leclercq and Rochais, in *Opera*, vol. 4, 460.
10 Cf. Rev 7:2. On the angel of great counsel, see above, chapt. 5, p. 16.
11 *Sententia communis* (Walther, *Proverbia*, no. 4360).
12 Heb 4:13.
13 Ps 9B:5.

*The divine eye's third [property of] sight produces tears of penitence, with the example of fire**.

Third, the divine gaze is something which produces tears of penitence. About this matter it is said in Revelation 19: "His eyes are like a flame of fire."[14] Now, a sinner is frozen by the frost of sin and has been made cold and hard as if he were ice. And therefore, just as the glare of fire melts ice, so God's gaze dissolves a sinner into tears. For this reason it is said in Luke 22: The Lord looked at Peter "and he went outside and wept bitterly."[15]

*The divine eye's fourth [property of] sight, which promotes hard work, with the example of a vineyard, and so on**.

Fourth, the divine gaze is something which promotes hard work, and for this reason Boethius says in book five of *The Consolation of Philosophy*: "If you do not wish to misrepresent it, a great necessity for honest behavior has been imposed on you since you do everything in the sight of a judge who perceives all things."[16] The Lord is like the head of a household who makes his laborers work all the harder because he observes them in the vineyard.[17] For he watches over the church as over a vineyard. And so he says of the vineyard of the church, "I, the Lord, am the one who preserves it," Isaiah 27.[18] And in the Psalm it is said: "The eyes of the Lord are on those who are righteous."[19] And not just any righteous people are spoken of here, but those who work at it, Romans 2: "Not the hearers of the law are the ones who are righteous before God, but the doers of the law will be justified."[20]

*The divine eye's fifth [property of] sight confers spiritual strength, with the example of the stars**.

Fifth, the divine gaze is something which confers spiritual strength, 2 Chronicles 16: "The eyes of the Lord contemplate the whole earth, and they give strength to those who believe in him with a righteous

14 Rev 19:12.
15 Luke 22:61.
16 Boethius, *Phil. cons.*, 5.pr.6.48, ed. Bieler, CCSL 94:105.
17 Cf. Matt 20:1–16.
18 Isa 27:3.
19 Ps 33:16.
20 Rom 2:13.

heart."[21] For if by its aspect a planet strengthens another star, either fixed or moving, so that it comes to have greater efficacy and power,[22] it is not surprising if the gaze of the most powerful and steadfast God gives power to the faint-hearted soul and increases the steadfastness and vigor of those who are not faint-hearted.

The divine eye's sixth [property of] sight heals the infirmities of the soul, with the example of the caladrius.

Sixth, the gaze of God's eyes is something which heals the infirmity of the soul, Ecclesiasticus 11: "One man is feeble," since he has been severely afflicted by the malady of a misdeed; "needing the restoration of his health," since he has been plundered of his spiritual goods; "lacking in strength," since he has been robbed of his powers of doing good; and "abounding in poverty," since he has been encumbered by a lack of spiritual goods; "and the eye of God has looked on him with good intentions," since by his gaze God has removed his malady and has restored the health he was missing.[23] For this reason the *Physiologus* says: "The *caladrius* is a bird with a white color. Its nature is said to be such that if it is brought to a sick person and closely observes the patient's face, this means he will live; but if it turns its eyes away from the sick person, this foretells his death."[24] And likewise, if Christ, pure in his way of life – as it were white in color – looks at a sinner with an eye of compassion, this not only means he will be healed, but it also brings it about, because if he were not to look at the sinner, the sinner would die for all time. Therefore, it is said in Ecclesiasticus 34: "The eye of the Lord is on those who fear him,"[25] and this phrase follows: "Giving health and life and blessing."[26]

21 2Chr 16:9.

22 "Aspect" in astrology refers to the angle between heavenly bodies in the chart of the zodiac, certain angles being thought to enhance the effects of the planets involved.

23 Sir 11:12–13.

24 *Physiologus*, 26 (V).2–3, 5–9, ed. Maurer, 89; cf. Hugh of Fouilloy (Ps.-Hugh of Saint-Victor), *Aviarium*, 2.53, ed. and trans. Clark, 228–31; Honorius of Autun, *Speculum ecclesiae*, De ascensione Domini, PL 172:958B–C.

25 Sir 34:19.

26 Sir 34:20.

The divine eye's seventh [property of] sight leads to the everlasting kingdom of glory.*

Seventh, the gaze of God's eyes is something that leads to the kingdom of glory, Job 36: God "does not withdraw his eyes from the righteous, and he places kings on the throne forever, and they are exalted there."[27] To that kingdom may he lead us, he who lives and reigns without end. Amen*.

27 Job 36:7.

Emended Readings and Significant Variations between *Pr1* and *P7P8*

Where not otherwise noted, entries here indicate material in *Pr1* that has had to be emended because it violates grammatical sense or unrecognizably alters the content of a passage in the best manuscript tradition of Peter's text. Other material in *Pr1* that, while not a breach of grammar, still significantly adds to, contracts, or expands what is found in the best manuscript tradition of Peter's text (represented by *P7P8*) is preceded here by the plus sign (+). The appendix is meant only to identify significant variations between *Pr1* and *P7P8* (or between *P7* and *P8* where they do not agree) and is not intended to be an exhaustive accounting. In the following, *om.* = "omitted from."

PROLOGUE

3: Here Begins the Moral Treatise on the Eye: Incipit tractatus moralis de oculo *P7P8*] Johannis Pithsani archiepiscopi Canthuariensis ordinis fratrum minorum liber de oculo morali foeliciter incipit. Prologus *Pr1*

CHAPTER 1

5: Alhacen: Alhacen *P7P8*] Alahacen *Pr1*; sclerotic: sclirotica *P7P8*] syrotica *Pr1*

6: + That seven things are necessary to protect the spiritual pupil: Quod septem sunt necessaria ad pupille spiritualis custodiam *Pr1*] *om. P7P8*

CHAPTER 2

8: + On the mercy and justice of prelates: De misericordia et iusticia prelatorum *Pr1*] *om. P7P8*; + And it is, thus, fitting that the chariot is spoken of first and then the horseman: Et ideo congrue primo currus, post auriga dictus est *Pr1*] *om. P7P8*

9: tension: intensio *P7P8*] intensione *Pr1*

10: fitting: decebit *P7P8*] licebit *Pr1*

CHAPTER 3

12: + On three types of vision among human beings: De triplici visione in homine *Pr1*] *om. P7P8*

CHAPTER 4

15: + It follows that two things are required for the forgiveness of someone who is sinful: Sequitur quod duo requiruntur ad iustificationem impij *Pr1*] *om. P7P8*

CHAPTER 5

16: + That one should beware of rash judgment: Quod iudicium temerarium sit cauendum *Pr1*] *om. P7P8*
17: very clean: mundissima *P7P8*] immundissima *Pr1*

CHAPTER 6

19: + That evil advisors and bribes corrupt prelates: Quod mali consiliarij et munera corrumpunt prelatos *Pr1*] *om. P7P8*
21: Brice: Bricius *P7P8*] Brichcius *Pr1*
23: Samuel: Samuelis *P7P8*] Heli *Pr1*
24: Martin: Martinus *P7P8*] Marcius alias Martinus *Pr1*; while the rule of law is scorned everyone will run: spreto moderamine iuris. Curritur *P7P8*] spreto moderamine iuris. Vnde versus. Curritur *Pr1*
25: Micah: Michee *P7P8*] Machabeorum *Pr1*
26: + On the examination of sin and its double punishment: De discussione peccati et duplicis pene *Pr1*] *om. P7P8*
27: extraordinary sight: spectaculum *P7P8*] speculum *Pr1*; + That prelates should beware of becoming a bad example or an enticement to sin: Quod prelatis cauendum sit a malo exemplo vel scandalo *Pr1*] *om. P7P8*
28: greater is the number of those: plurium *P7P8*] plurimum *Pr1*
29: + From which it follows that sins of the flesh are of greater disrepute than spiritual sins: Ex quo sequitur quod peccata carnalia maioria [*sic*; read: maioris] infamie sunt quam spiritualia *Pr1*] *om. P7P8*; and moved: sumaturque *P7P8*] sumatur *Pr1*; + From which it is educed that Holy Scriptures are seen directly by the person who fulfills them in his deeds: Ex quo elicitur, quod sacram scripturam directe videt, qui eam opere complet *Pr1*] *om. P7P8*
31: + On the truth of a trinity and unity: De trinitatis et vnitatis veritate *Pr1*] *om. P7P8*; optical phenomenon: spectaculum *P7P8*] speculum *Pr1*

32: + That God is a mirror without blemish: Quod deus est speculum sine macula *Pr1*] *om. P7P8*

33: + from which it follows that encumbered by our body, we cannot see the divine essence: vnde sequitur quod corpore pregrauati diuinam essentiam videre non possumus *Pr1*] *om. P7P8*

34: + from which it is proven that what the world considers favorable is unfavorable in God's sight, and vice versa: vnde probatur quod mundus reputat prosperum coram deo esse aduersum et econuerso *Pr1*] *om. P7P8*

35: of God: Dei *P7P8*] *om. Pr1*; crossing: cancellacione *P7P8*] collatione *Pr1*

37: + That knowing how to differentiate sins is necessary for prelates: Quod scientia discernendi peccata prelatis sit necessaria *Pr1*] *om. P7P8*

38: + from which it is deduced that the judgment of a person's sight is deceived concerning the rich and the poor: Ex quo trahitur quod fallatur iudicium visus hominis circa pauperes et diuites *Pr1*] *om. P7P8*

41: + And on preserving humility in a lofty position: Et de seruanda humilitate in sublimitate *Pr1*] *om. P7P8*; Alhacen: Alahacen *P7P8Pr1*

CHAPTER 7

45: + On the value of the communal life: De bono socialis vite *Pr1*] Prima proprietas *P7*, *om. P8*; Alhacen: Alhacen *P7P8*] Alachacen *Pr1*

46: coenobitic: cenobiticam *P8*] cenobicam *Pr1*, *om. P7*

47: + On the second moral instruction, namely on fools saying many things and doing few, knowing a great deal and loving only a little: De secunda instructione morali, scilicet de stultis multa dicentibus et pauca facientibus, multum cognoscentibus et parum diligentibus *Pr1*] Secunda proprietas *P7P8*

50: + The third moral instruction. On maintaining the memory of death, which worldly people hate: Tercia instructio moralis. De memoria mortis habenda quam mundiales despiciunt *Pr1*] Tercia proprietas *P7P8*

53: + The fourth moral instruction. On preserving a singularity of intention: Quarta instructio moralis. De simplicitate intentionis seruanda *Pr1*] Quarta proprietas *P7P8*; many: multi *P7P8*] multa *Pr1*

54: + The fifth lesson. On avoiding vainglory: Quinta informatio. De vitanda inani gloria *Pr1*] Quinta proprietas *P7P8*

57: + The sixth moral lesson. That the eyes, that is of the intellect and of the affections, should be turned to God: Sexta informatio moralis. Quod oculi, id est intellectus et affectus, dirigendi sunt ad deum *Pr1*]

Sexta proprietas *P7P8*; 2 Chronicles 20: ii Paralippomenon xx *P7P8*] i Paralippomenon x *Pr1*; 27: xxvi *P7P8*, xxiij *Pr1*

58: *Policraticus*: Policraticus *P8*] Policratus *P7Pr1*

59: + The seventh instruction. That one should visualize Christ's wounds in one's memory: Septima instructio. Quod vulnera cristi per memoriam sunt aspicienda *Pr1*] Septima proprietas *P7P8*

61: and at: atque ad *P7P8*] ad que *Pr1*; + The eighth moral instruction. Against fools neglecting their own deeds and rashly judging others' deeds: Octaua instructio moralis. Contra fatuos sua negligentes et aliorum facta temere iudicantes *Pr1*] Octaua proprietas *P7P8*

62: Why: Quid *P7P8*] Quod *Pr1*

63: illuminated: illustrati *P7P8*] illustra *Pr1*; Scythia: Scicia *P7P8*] sicilia *Pr1*

64: 17: xvii *P7P8*] xviij *Pr1*; Sometimes: Quandoque *P7P8*] Quoniam *Pr1*

65: + The ninth moral instruction. That sinners look down on heavenly matters and esteem worldly matters: Nona instructio moralis. Quod peccatores celestia paruipendunt, et terrena magnipendunt *Pr1*] Nona proprietas *P7P8*

67: + The tenth moral instruction. That we should freely share with others the goods that we possess: Decima instructio moralis. Quod bona, que habemus, alijs libenter communicare debemus *Pr1*] Decima proprietas *P7P8*

69: Lucilius: Lucilium] Lucillum *Pr1*, *om. P7P8*

70: + If wisdom were given to me with the stipulation that I keep it to myself, I would not scorn it, I would abandon it. In fact, there is no pleasure in possessing something good without a companion: Si cum hac exceptione detur sapientia, vt intus teneam non reijciam, abijciam. Nullius siquidem boni sine socio iocunda est possessio *Pr1*] *om. P7P8*; Dindimus: Dindimum *P7P8*] Didimum *Pr1*; 82: lxxxii *P7P8*] lxxxiij *Pr1*

71: + The eleventh moral instruction. That the state of someone's mind is signaled through his body's eyes: Undecima instructio moralis. Quod per oculum corporis perpenditur status mentis *Pr1*] Vndecima proprietas *P7P8*; 12: xii *P7P8*] ij *Pr1*; and: et *P7P8*] *om. Pr1*

72: wandering: uagus *P7P8*] fagus *Pr1*; + The twelfth moral instruction. That it is sometimes useful to leave external matters behind and return to internal ones: Duodecima instructio moralis. Quod vtile sit relictis exterioribus aliquando ad interiora conuerti *Pr1*] Duodecima proprietas *P7P8*; internal: interioribus *P7P8*] exterioribus *Pr1*

73: 78: lxxviii *P7P8*] xxvij *Pr1*

CHAPTER 8

75: + Vices: viciorum *Pr1*] uiciorum. Et primo de oculo sublimi et tumido (inuido *P7*) superborum *P7P8*; arrogant: tumidi *P7P8*] timidi *Pr1*

78: + In a letter to King Alexander, Dindimus, king of the Brahmins, wrote as follows: "We are warmed by the sun, we receive moisture from the dew, we quench our thirst from a brook, the ground serves as our bed, anxiety does not interrupt our sleep, thinking does not weary our mind. We have no grand designs to rule over human beings like ourselves, nor do we demand servitude from anything besides our bodies, which we value only as the servants to our souls. And besides, it is an act of barbarity to force people into compliance when the same nature gave birth to them as to us as brothers, just as they, too, are promised an inheritance from the one God, the father of all good things": Didimus [*sic*] rex bragmanorum in epistola quadam ad allexandrum regem sic scripsit: Sole calescimus, rore humectamur, sitim riuo frangimus, thorum ministrat humus, somnum solicitudo non rumpit, mentem cogitatio non fatigat. In homines nobis similes supra non agitamus imperium, nec a quolibet seruitium exigimus preterquam a corpore, quod solum anime famulari censemus. Alioquin seuicia est in obsequium cogere, quos nobis fratres eadem natura genuit, vt quibus ab vno deo patre omnium bonorum spondetur hereditas *Pr1*] *om. P7P8*

79: + On the second distinction, namely envy. On the unhappy eye of the envious person pining away at the good fortune of his fellow human being: De secunda differentia scilicet inuidia. De infelici oculo inuidi thabescentis de prosperitate proximi *Pr1*] Secundo de oculo liuenti et strabonico inuidorum *P7P8*

82: grief: tristicia *P7P8*] iusticia *Pr1*; the envious: inuidos *P7P8*] inuidus *Pr1*; + On the third distinction, namely wrath. On the agitated eye of the angry man, who is temporarily insane: De tercia differentia scilicet Ira. De turbato oculo iracundi ad tempus insani *Pr1*] Tercio de oculo turbato iracundorum *P7P8*

83: Architrenius: Architrenius *P8*] Architrinius *P7*, archita tharentinus *Pr1*; infectious: infectiui *P7P8*] ignei *Pr1*

84: Salvina: Saluiam *P7P8Pr1*

85: would deserve: mereretur *P8*] meretur *P7*, moreretur *Pr1*; 36: xxxvi *P7P8*] xxxi *Pr1*

86: ought to be angry: irascatur *P7P8*] irascitur *Pr1*; + On the fourth distinction, namely spiritual sloth. On the putrid eye of the slothful person, lazy and drowsy when it comes to his own salvation: De quarta differentia, scilicet accidia. De putrido oculo accidiosi ad salutem suam pigri et somnolenti *Pr1*] Quarto de oculo sompnolento accidiosorum *P7P8*

88: a story of this very kind: narracionem huiuscemodi *P7P8*] narrationis huius *Pr1*

91: + The fifth distinction follows, namely avarice. On the eye of the avaricious person blinded by earthly possessions: Sequitur quinta differentia scilicet auaritia. De oculo auari terrenis possessionibus excecati *Pr1*] Quinto de oculo puluerulento auarorum *P7P8*

93: years: anni *P7P8*] amici *Pr1*; Petronius: Petronius *P7P8*] Petranius *Pr1*

94: Whoever considers with what a great heart Christ loves us preserves his heart, and whoever understands how great was the bitterness of Christ's passion preserves his gall bladder: Cor eius reponit, qui ex quanto corde Christus dilexerit (dilexit *P7*) nos attendit; fel eciam reponit, qui quanta fuerit eius amaritudo passionis intelligit *P7P8*] cor eius reponit, qui quanta fuerit eius amaritudo passionis intelligit *Pr1* (by homoteleuton); + On the sixth distinction, namely gluttony. On the eye of the gluttonous person wounded by being excessively voracious: De sexta differentia scilicet gula. De oculo gulosi nimia ingluuie sauciati *Pr1*] Sexto de oculo carneo gulosorum *P7P8*

96: + Not: hastily, too expensively, excessively, too eagerly, with too much care: Non propere, laute, nimis, ardenter, studiose *Pr1*] *om.* *P7P8*

97: undermining: suffossio *P8*, suffosio *P7*] confossio *Pr1*

98: + The seventh distinction, on lust. That a woman's eyes are the darts of unchasteness by which many are wounded: Septima differentia de luxuria. Quod oculi mulieris sint tela impudicicie quibus multi vulnerantur *Pr1*] Septimo de oculo impudico luxuriosorum *P7P8*

102: + Conversation with them is nothing but bitter venom: Colloquium quarum non est nisi virtus [*sic; in source text*: virus] amarum *Pr1*] *om.* *P7P8*

CHAPTER 9

108: books: librorum *P7P8*] liberum *Pr1*

109: to counterfeit: mentiri *P7P8*] metri *Pr1*; 1092: M°xcii *P7P8*] milesimoduodecimo *Pr1*; cheaper: uiliores *P7P8*] meliores *Pr1*

110: 3: iii *P7P8*] iiij *Pr1*

CHAPTER 10

111: every: omnem *P7P8*] non omnem *Pr1*; *Policraticus*: Policraticus *P7P8*] pollicratus *Pr1*

112: Thérouanne: Teruanensis *P7P8*] carnotensis *Pr1*

113: Claudius: Claudius *P7P8*] Claudianus *Pr1*; *Attic*: Atticarum *P8*] articatum *P7*, articarum *Pr1*

CHAPTER II

114: + The first lesson. On humility in learning, being scornful of no one, and enduring insults: Prima informatio. De humilitate discendi, nullum contemnendi et iniurias tolerandi *Pr1*] *om. P7P8*

115: was shown: monstrata est *P7P8*] monstrantem *Pr1*

116: could teach: doceatur *P7P8*] docetur *Pr1*

118: bends downward: demittit *P7P8*] dimittit *Pr1*

120: your: uestri *P7P8*] vestre *Pr1*; + On the love of poverty among students: De amore paupertatis in studentibus *Pr1*] *om. P7P8*; of reason: racionis *P7P8*] oculorum *Pr1*; *Policraticus*: Policraticus *P7P8*] Policratus *Pr1*

121: Vitruvius: uictruinus *P8*, uictrinius *P7*, victorius *Pr1*

122: + On the chastity and abstinence of students: De castitate et abstinentia studentium *Pr1*] *om. P7P8*

123: wisdom: Sapiencia *P7P8*] Sapientie *Pr1*; Ahimelech: Achimelech *P8*] Archimelech *P7*, Abimalech *Pr1*

125: + It follows next that scholars should entrust their books not so much to a book bag as to their memory: Sequitur de libris scolarium non tam sacculo quam memorie commendandis *Pr1*] Secunda condicio *P7*, *om. P8*; 10: x *P7P8*] vicesimo *Pr1*; breeze: aure *P7P8*] aures *Pr1*

127: + That a separation from God through students' sinning destroys a knowledge of divine matters: Quod elongatio a deo per peccatum a studentibus tollit noticiam diuinorum *Pr1*] Tercia condicio *P7*, *om. P8*

128: affections: affectu *P7P8*] effectu *Pr1*

129: + The fourth lesson. That one should be zealous about the kinds of knowledge that have substance, not those that are baseless, and so on: Quarta informatio. Quod in scientijs solidis studendum sit non vanis et cetera *Pr1*] Quarta condicio *P7*, *om. P8*

130: surely: profecto *P7P8*] perfecto *Pr1*; something: queque *P7P8*] quantum *Pr1*

132: unfailing: indeficientis *P7P8*] indeficiens *Pr1*; Cluny: Cluniacensis *P7P8*] damasceni *Pr1*

133: obscenities: obcenitatibus *P7P8*] obstematibus *Pr1*

134: Epimenides: Eumenidis *P7P8Pr1*; + The fifth lesson, on diligent attention to and retention of what is taught, and the indefatigable love for the devotion to learning: Quinta informatio de diligenti attentione et retentione doctrine, et indefesso circa studium discendi amore *Pr1*] Quinta condicio *P7*, *om. P8*

136: Machetes: Machetem *P7P8*] machatem *Pr1*

137: Solon: Solon *P8*] Salomon *P7*, Nausolon *Pr1*

138: right hand: dextreque *P7P8*] dextroque *Pr1*

139: to slumber: sopori *P8*] corpori *P7*, soporis *Pr1*; with sleep: sompno *P7P8*] somnum *Pr1*

140: + Note that there are seven keys to being a philosopher, which the *Policraticus* sets out in these verses: A humble mind, a passion for investigating things, a settled life, / A quiet study, poverty, dwelling off the beaten track. And the seventh key is affection for one's teachers. For this reason, Quintilian also says: Teachers are accustomed to making those difficult passages that must be read accessible to many: Nota septem sunt claues philosophandi, quos ponit policratus [*sic*] istis versibus: Mens humilis, studium querendi, vita quieta. Scrutinium tacitum, paupertas, terra remota. Et septima clauis est amor docentium. Vnde et quintillianus ait: Hec reserare solent multis obscura legenda *Pr1*] *om. P7P8*; + The sixth lesson. That one must zealously dwell on reading, nor should one leave off reading before death: Sexta informatio. Quod studio lectionis sit insistendum, nec ante mortem desistendum *Pr1*] Sexta conditio *P7*, *om. P8*; second: secundo *P7P8*] quinto *Pr1*

143: + The seventh instruction. That students lack the light of grace, which is God. And on the threefold eye, and so on: Septima instructio. Quod studentes indigent lumine gratie, que deus est. Et de triplici oculo et cetera *Pr1*] Septima condicio *P7*, *om. P8*

144: + This is why the Apostle says, "His anointing teaches us in all things": Vnde dicit apostolus: Vnctio eius docet nos de omnibus *Pr1*] *om. P7P8*; Bernard: Bernardus *P7P8*] Gregorius *Pr1*

CHAPTER 12

146: + The first instruction for prelates. That just as prelates surpass others in their position, so should they in holiness: Prima instructio prelatorum. Quod prelati sicut dignitate sic precellere debent sanctitate *Pr1*] *om. P7P8*

147: you have been made: factus es *P7P8*] factus est *Pr1*

149: + The second instruction for prelates. That it is the duty of a watchman, that is to say a prelate, to keep watch and not to be blind: Secunda instructio prelatorum. Quod speculatoris, id est prelati, est vigilare non cecum esse *Pr1*] De secunda instructione *P7*, *om. P8*; 77: lxxvii *P7P8*] lxxviij *Pr1*

150: not wanting: nolentes *P7P8*] volentes *Pr1*

152: 34: xxxiiii *P7P8*] xxiiii *Pr1*

153: + The third instruction for prelates. That a prelate should carefully scrutinize not only others' behavior, but also the behavior of those close to him: Tercia instructio prelatorum. Quod prelatus debet non solum aliorum sed etiam suorum mores diligenter discutere *Pr1*] De tercia instructione *P7*, *om. P8*; Sabinian: fabinianum *P7P8*, fabianum *Pr1*; would see: Videret *P7P8*] Vident *Pr1*

154: allow: sinit *P7P8*] sunt *Pr1*

155: Edmund: Edmundo *P7*] Eadmundo *P8*, Odimundo *Pr1*; evil intentions: animi *P7P8*] agni *Pr1*

157: + The fourth instruction for prelates. That prelates and clergymen, whose eyes are weakened, are worse than the laity: Quarta instructio prelatorum. Quod prelati et clerici caligantibus oculis peiores sunt laycis *Pr1*] De quarta instructione *P7*, *om. P8*

158: Sens: Senonensis *P7P8*] senacensis *Pr1*; to: ut *P7P8*] et *Pr1*

159: name: nomen *P7P8*] non *Pr1*; + The fifth instruction for prelates, on prelates' dutiful compassion for their congregants, and on the ostrich egg: Quinta instructio prelatorum de pia compassione prelatorum erga subditos, et de ouo strutionis *Pr1*] De quinta instructione *P7*, *om. P8*

160: 5: v *P7P8*] vi *Pr1*; 11: xi *P7P8*] xij *Pr1*

161: Frontinus: froncius *P7P8*] francius *Pr1*; opponents: aduerse parti *P7P8*] aduerso parti *Pr1*; brushwood: sarmenta *P7P8*] sacrum *Pr1*

162: + The sixth instruction for prelates, on the moderation prelates should preserve in temporal matters: Sexta instructio prelatorum. De mediocritate prelatorum seruanda circa temporalia *Pr1*] De sexta instructione *P7*, *om. P8*

163: of the church: ecclesiastici *P7P8*] ecclesiasti *Pr1*

165: he ... gapes: inhiat *P7P8*] inhiant *Pr1*

166: + The seventh instruction for prelates: that sin is much more unsightly and harmful in a priest or a prelate than in a member of the laity: Septima instructio prelatorum, quod multo deformius et nociuius est peccatum in sacerdote aut prelato quam laico *Pr1*] De septima instructione *P7*, *om. P8*

168: church's: ecclesie *P7P8*] *om. Pr1*; Silvestris: Siluestrem *P7P8*] siluestrum *Pr1*

169: 3: iii *P7P8*] ij *Pr1*

CHAPTER 13

171: + On paying attention to one's own sin: De consideratione proprij peccati *Pr1*] *om. P7P8*; now: nunc *P7P8*] non *Pr1*

172: + On paying attention to the misery of one's fellow human being: De consideratione miserie proximi *Pr1*] Secundum quod oculus debet respicere *P7*, *om. P8*

173: a man: unus *P7P8*] *om. Pr1*

175: + Then follows paying attention to punishment in hell: Sequitur consideratio pene yehennalis *Pr1*] Tercium quod debet oculus respicere *P7*, *om. P8*
176: as well as: tantum] tantam *Pr1*, *om. P7P8*
177: something sharply bitter: mordicatiuum *P7P8*] medicamentum *Pr1*; + On paying attention to life in heaven: De consideratione vite celestis *Pr1*] Quartum quod debet oculus respicere *P7*, *om. P8*
179: timbrel: tympanum *P7P8*] tympalium *Pr1*

CHAPTER 14

182: placed over: superpositam *P7P8*] suppositam *Pr1*; me: me *P7P8*] *om. Pr1*; + The second thing that delights vision signifies that one should carefully consider the verdant examples of the saints: Secundum quod delectat visum significat quod virentia sanctorum exempla sunt diligenter attendenda *Pr1*] Secundum delectabile *P7*, *om. P8*
184: way of life: conuersacionis *P7P8*] conuersionis *Pr1*
186: On the third thing that delights vision. That we should frequently look in a mirror, i.e., Jesus Christ: De tercio quod delectat visum. Quod speculum, id est ihesum cristum frequenter respicere debemus *Pr1*] Tercium delectabile *P7*, *om. P8*; cleanness: mundicie *P7P8*] immundicie *Pr1*

CHAPTER 15

189: + That the divine eye looks at everything: quod diuinus oculus omnia respicit *Pr1*] *om. P7P8*; 23: xxiii *P7P8*] xxiiij *Pr1*; Damiani: Damianus *P7P8*] damascenus *Pr1*; barber: tonsorem *P7P8*] tonsuram *Pr1*
190: + The divine eye's second [property of] sight, by which it inspires awe, with the example of the sun: Secundus intuitus diuini oculi, quo incutit timorem solis exemplo *Pr1*] *om. P7P8*; of judgment: iudicii *P7P8*] in iudicij *Pr1*
191: + The divine eye's third [property of] sight produces tears of penitence, with the example of fire: Tercius intuitus diuini oculi lacrimas penitentiales eduxit exemplo ignis *Pr1*] *om. P7P8*; + The divine eye's fourth [property of] sight, which promotes hard work, with the example of a vineyard, and so on: Quartus intuitus diuini oculi, qui operationes strenuas promouet exemplo vinee etcetera *Pr1*] *om. P7P8*; + The divine eye's fifth [property of] sight confers spiritual strength, with the example of the stars: Quintus diuini oculi intuitus spiritualem fortitudinem confert exemplo stellarum *Pr1*] *om. P7P8*

192: + The divine eye's sixth [property of] sight heals the infirmities of the soul, with the example of the *caladrius*: Sextus intuitus diuini oculi infirmitates anime sanat exemplo caladrij *Pr1*] *om. P7P8*

193: + The divine eye's seventh [property of] sight leads to the everlasting kingdom of glory: Septimus intuitus ad regnum glorie sempiternum perducit *Pr1*] *om. P7P8*; Amen: Amen *P8*] Amen. Explicit liber de oculo morali. *P7*, Amen. Tractatus (Johannis Pithsani archiepiscopi Cantuariensis) de oculo morali finit feliciter *Pr1*

Bibliography

Manuscripts

Assisi, Convento dei Frati Minori MS. Com. 442
Basel, Universitätsbibliothek MS. B X 34
Bruges, Stadsbibliotheek MS. 167
Durham, University Library MS. Mickleton & Spearman 89
London, British Library MS. Harley 219
London, Lambeth Palace Library MS. 483
Oxford, Bodleian Library MS. Bodl. 315 (*SC* 2712)
Oxford, Bodleian Library MS. Bodl. 742 (*SC* 2756)
Oxford, Bodleian Library MS. Hamilton 21 (*SC* 24451)
Oxford, Bodleian Library MS. Laud misc. 706
Paris, Bibliothèque Mazarine MS. 794 (1244)
Paris, Bibliothèque Mazarine MS. 888 (1136)
Paris, Bibliothèque nationale de Paris MS. fr. 24402
Paris, Bibliothèque nationale de Paris MS. lat. 7434
Paris, Bibliothèque nationale de Paris MS. lat. 15559
Paris, Bibliothèque nationale de Paris MS. lat. 15588
Paris, Bibliothèque nationale de Paris MS. lat. 15863
Paris, Bibliothèque nationale de Paris MS. lat. 16112
Paris, Bibliothèque nationale de Paris MS. lat. 16113
Paris, Bibliothèque nationale de Paris MS. lat. 16114
Paris, Bibliothèque nationale de Paris MS. lat. 16119
Paris, Bibliothèque nationale de Paris MS. lat. 16122
Paris, Bibliothèque nationale de Paris MS. lat. 16157
Paris, Bibliothèque nationale de Paris MS. lat. 16390
Paris, Bibliothèque nationale de Paris MS. lat. 16395
Paris, Bibliothèque nationale de Paris MS. lat. 16396
Paris, Bibliothèque nationale de Paris MS. lat. 16407
Paris, Bibliothèque nationale de Paris MS. lat. 16616
Paris, Bibliothèque nationale de Paris MS. lat. 16681
Sankt Florian, Augustiner-Chorherrenstift MS. XI.130
Sankt Florian, Augustiner-Chorherrenstift MS. XI.164
Sigmaringen, Fürstlich Hohenzollern'sche Bibliothek MS. 12
Trier, Stadtbibliothek MS. 574/1522 8°
Uppsala, Universitetsbibliothek MS. C 641
Würzburg, Universitätsbibliothek MS. I.t.f.201

Primary Sources

Adam Scotus. *Sermones*. PL 198:91–440.

Alan of Lille. *Anticlaudianus*. Ed. Robert Bossuat. Textes philosophiques du moyen âge, 1. Paris: Vrin, 1955.

—. *Anticlaudianus; or The Good and Perfect Man*. Trans. James J. Sheridan. MST 14. Toronto: Pontifical Institute of Mediaeval Studies, 1973.

—. *De fide catholica contra haereticos libri iv*. PL 210:305A–430A.

—. *Sermo de sphaera intelligibili*. Ed. Marie-Thérèse d'Alverny. In *Textes inédits*. Études de philosophie médiévale, 52. Paris: Vrin, 1965. Pp. 295–306.

—. *Theologicae regulae*. PL 210:621A–84C.

—. *Règles de théologie; suivi de Sermon sur la sphère intelligible*. French trans. Françoise Hudry. Paris: Cerf, 1995.

Albert the Great. *Summa theologiae. Pars prima*. Ed. Auguste Borgnet. In *Opera omnia*. 38 vols. Paris: Vivès, 1890–1895. Vol. 31.

Ps.-Albert the Great (Adenulf of Anagnia [?]). *Commentarii in Psalmos*. Ed. Auguste Borgnet. 3 vols. In *Opera omnia*. 38 vols. Paris: Vivès, 1890–1895. Vol. 15–17.

—. *Commentary on the Psalms*. Trans. Joseph Kenny. Available at http://www.josephkenny.joyeurs.com/CDtexts/PsalmsAlbert/default.htm (accessed November 12, 2011).

Aldhelm of Malmesbury. *De virginitate*. Ed. Rudolf Ehwald. In *Opera*. MGH, Auctores antiquissimi, 15. Berlin: Weidmann, 1919; reprint 1961. Pp. 226–471.

—. *The Prose Part of "De virginitate."* In *The Prose Works*. Trans. Michael Lapidge and Michael Herren. Cambridge, Engl.: D.S. Brewer, 1979; reprint Woodbridge, Engl., and Rochester, NY: D.S. Brewer, 2009.

Alexander IV (Pope). *Les registres d'Alexandre IV; recueil des bulles de ce pape publiées ou analysées d'après les manuscrits originaux des archives du Vatican*. Ed. Charles Bourel de La Roncière, Jean de Loye, and Auguste Coulon. 3 vols. Paris: A. Fontemoing, 1902–1959.

Alexander of Canterbury. *Liber ex dictis beati Anselmi:* Ed. Richard W. Southern and Franciscus S. Schmitt. In *Memorials of St. Anselm*. Auctores Britannici Medii Aevi, 1. Oxford: Oxford University Press, 1969. Pp. 105–195.

Alexander of Hales (and his students). *Summa theologica*. Ed. PP. Collegii s. Bonaventurae. 4 vols. Quaracchi: ex typographia Collegii s. Bonaventurae, 1924–1948.

Alhacen, *De aspectibus*, 1–3. Ed. and trans. A. Mark Smith. In *Alhacen's Theory of Visual Perception. A Critical Edition, with English Translation and Commentary, of the First Three Books of Alhacen's De Aspectibus, the Medieval Latin Version of Ibn al-Haytham's Kitab al-Manazir*. 2 vols. Transactions of the American Philosophical Society, 91, Parts 4–5. Philadelphia: American Philosophical Society, 2001.

—. *De aspectibus*, 4–5. Ed. and trans. A. Mark Smith. In *Alhacen on the Principles of Reflection. A Critical Edition, with English Translation and Commentary, of Books*

4 and 5 of Alhacen's De Aspectibus, the Medieval Latin Version of Ibn al-Haytham's Kitab al-Manazir. 2 vols. Transactions of the American Philosophical Society, 96, Parts 2–3. Philadelphia: American Philosophical Society, 2006.

—. *De aspectibus*, 6. Ed. and trans. A. Mark Smith. In *Alhacen on Image-Formation and Distortion in Mirrors. A Critical Edition, with English Translation and Commentary, of Book 6 of Alhacen's De Aspectibus, the Medieval Latin Version of Ibn al-Haytham's Kitab al-Manazir*. 2 vols. Transactions of the American Philosophical Society, 98, Part 1, Sections 1–2. Philadelphia: American Philosophical Society, 2008.

—. *De aspectibus*, 7. Ed. and trans. A. Mark Smith. In *Alhacen on Refraction. A Critical Edition, with English Translation and Commentary, of Book 7 of Alhacen's De Aspectibus, the Medieval Latin Version of Ibn al-Haytham's Kitab al-Manazir*. 2 vols. Transactions of the American Philosophical Society, 100, Part 3, sections 1–2. Philadelphia: American Philosophical Society, 2010.

—. *Opticae*. Ed. Federicus Risner. In *Opticae thesaurus Alhazeni arabis libri septem, nunc primum editi. Eiusdem liber De crepusculis et nubium ascensionibus. Item Vitellonis Thuringopoloni libri X. Omnes instaurati, figuris illustrati et aucti, adiectis etiam in Alhazenum commentarijs*. Basel: per Episcopios, 1572; reprint New York: Johnson Reprint Corp., 1972. Pp. 1–282.

Ambrose of Milan. *Exameron*. Ed. Karl Schenkl. *Opera*. Part 1. CSEL 32.1. Vienna, etc.: Tempsky, 1897. Pp. 1–261.

—. *Hexameron, Paradise, and Cain and Abel*. Trans. John J. Savage. FOTC 42. New York: Fathers of the Church, 1961.

—. *De virginibus*. Ed. and trans. Franco Gori. In *Verginità e vedovanza, I*. Sancti Ambrosii Episcopi Mediolanensis Opera, 14/1. Opere Morali 2/1. Milan: Biblioteca Ambrosiana; Rome: Città Nuova Editrice, 1989. Pp. 99–241.

—. *On Virgins*. Trans. Boniface Ramsey. In *Ambrose*. Early Church Fathers. London and New York: Routledge, 1997. Pp. 71–116.

—. *Concerning Virgins*. Trans. H. de Romestin, E. de Romestin, and H.T.F. Duckworth. NPNF2, 10.

Anselm of Canterbury. *Orationes*. Ed. Franciscus S. Schmitt. In *Opera omnia*. 6 vols. Edinburgh: T. Nelson and Sons, 1946–1961. Vol. 3. Pp. 5–75.

—. *The Prayers and Meditations of St. Anselm with the Proslogion*. Trans. Sister Benedicta Ward. Harmondsworth: Penguin, 1973.

Ps.-Anselm of Canterbury. *Liber Anselmi Archiepiscopi de humanis moribus per similitudines*. Ed. Richard W. Southern and Franciscus S. Schmitt. In *Memorials of St. Anselm*. Auctores Britannici Medii Aevi, 1. Oxford: Oxford University Press, 1969. Pp. 37–104.

—. *Liber de similitudinibus*. PL 159:605A–708D.

Aristotle. *De anima*. Ed. David Ross. Oxford: Clarendon, 1961.

—. *De anima*. Trans. John A. Smith. In *The Works of Aristotle Translated into English*. Ed. William D. Ross. 8 vols. Oxford: Clarendon, 1931; reprint 1963. Vol. 3.

—. *De generatione animalium*. Trans. William of Moerbeke. Ed. Hendrik J. Drossaart Lulofs. Aristoteles Latinus, 17.2.5. Corpus Philosophorum Medii Aevi, 1.17.2.5. Bruges and Paris: Desclée de Brouwer, 1966.

—. *De generatione animalium*. Trans. Michael Scot. Ed. Aafke M.I. van Oppenraaij. In *De animalibus. Michael Scot's Arabic-Latin Translation. Part Three: Books XV–XIX: Generation of Animals*. Aristoteles Semitico-latinus, 5. Leiden, New York, and Cologne: Brill, 1992.

—. *Generation of Animals*. Trans. Arthur L. Peck. Rev. ed. LCL 366. Cambridge, MA: Harvard University Press; London: W. Heinemann, 1953.

—. *Historia animalium*. Vol. 1: Books I–X: Text. Ed. David M. Balme. Cambridge Classical Texts and Commentaries, 38. Cambridge, Engl.: Cambridge University Press, 2002.

—. *De historia animalium*. Trans. William of Moerbeke. Ed. Pieter Beullens and Fernand Bossier. Pars 1: Lib. 1–5. Aristoteles Latinus, 17.2.1.1. Leiden, etc.: Brill, 2000.

—. *History of Animals*. Trans. Arthur L. Peck and David M. Balme. 3 vols. LCL 437–439. Cambridge, MA: Harvard University Press; London: W. Heinemann, 1965–1991.

—. *De insomniis*. Ed. and trans. David Gallop. In *Aristotle on Sleep and Dreams. A Text and Translation with Introduction, Notes and Glossary*. Warminster, Engl.: Aris & Phillips Ltd., 1996.

—. *Metaphysica*. Translatio anonyma sive "Media." Ed. Gudrun Vuillemin-Diem. Aristoteles Latinus, 25.2. Corpus Philosophorum Medii Aevi, 1.25.2. Leiden: Brill, 1976.

—. *Metaphysica*. Trans. William of Moerbeke. Ed. Gudrun Vuillemin-Diem. 2 vols. Aristoteles Latinus, 25.3.1–2. Corpus Philosophorum Medii Aevi, 1.25.3.1–2. Leiden, etc.: Brill, 1995.

—. *Metaphysics*. Trans. Hugh Tredennick. 2 vols. LCL 271, 287. Cambridge, MA: Harvard University Press, 1933–1935; reprint 1969–1975.

—. *Ethica Nicomachea*. Trans. Robert Grosseteste (textus recognitus and textus purus). Ed. René Antoine Gauthier. In *Ethica Nicomachea*. 5 vols. Aristoteles Latinus, 26.1–3. Fasc. 1–5. Corpus Philosophorum Medii Aevi, 1.26.1–3/1–5. Leiden: Brill; Brussels: Desclée de Brouwer, 1972–1974. Fasc. 3–4.

—. *The Nicomachean Ethics*. Trans. Harris Rackham. New and rev. ed. LCL 73. Cambridge, MA: Harvard University Press; London: W. Heinemann, 1934; reprint 1975.

—. *De partibus animalium libri iv*. Ed. Bernhard Langkavel. Aristotelis opera, 1. Leipzig: Teubner, 1868.

—. *Parts of Animals. Movement of Animals. Progression of Animals*. Trans. Arthur L. Peck and Edward S. Forster. LCL 323. Cambridge, MA: Harvard University Press; London: W. Heinemann, 1937; reprint 1968.

—. *Topica*. Trans. Boethius (recensio alpha). Ed. Lorenzo Minio-Paluello. Aristoteles Latinus, 5.1–3. Corpus Philosophorum Medii Aevi, 1.5.1–3. Bruges: Desclée de Brouwer, 1969.

—. *Posterior Analytics. Topica.* Trans. Hugh Tredennick and Edward S. Forster. LCL 391. Cambridge, MA: Harvard University Press, 1960; reprint 1976.

Ps.-Aristotle. *Secretum secretorum.* Ed. Robert Steele. In *Opera hactenus inedita Rogeri Baconi ...* Part 5: *Secretum secretorum cum glossis et notulis ... accedunt versio Anglicana ex Arabico edita per A. S. Fulton, versio vetusta Anglo-Normanica nunc primum edita.* Oxford: Clarendon, 1920. Pp. 1–175.

Augustine of Hippo. *De agone christiano.* Ed. Joseph Zycha. In *De fide et symbolo; De fide et operibus; De agone christiano; De continentia.* CSEL 41. Vienna: Tempsky, 1900.

—. *Christian Instruction. Admonition and Grace. The Christian Combat. Faith, Hope and Charity.* Trans. John J. Gavigan, John Courtney Murray, Robert P. Russell, and Bernard M. Peebles. FOTC 2. 2nd ed. Washington, DC: Catholic University of America Press, 1950.

—. *De civitate Dei.* Ed. Bernhard Dombart and Alfons Kalb. 2 vols. CCSL 47–48. Turnhout: Brepols, 1955.

—. *The City of God.* Trans. Demetrius B. Zema and Gerald G. Walsh. 3 vols. FOTC 8, 14, 24. Washington, DC: Catholic University of America Press, 1950–1977.

—. *The City of God.* Trans. Marcus Dods. NPNF1, 2.

—. *Confessiones.* Ed. Luc Verheijen. CCSL 27. Turnhout: Brepols, 1981.

—. *The Confessions.* Trans. Maria Boulding. WSA, Part 1, vol. 1. Hyde Park, NY: New City Press, 1997.

—. *The Confessions.* Trans. Joseph G. Pilkington. NPNF1, 1.

—. *Contra academicos libri tres.* Ed. William M. Green. In *Contra academicos; De beata vita; necnon De ordine libri.* CCSL 29. Turnhout: Brepols, 1970. Pp. 1–61.

—. *Against the Academics.* Trans. John J. O'Meara. Ancient Christian Writers, 12. Westminster, MD: Newman Press, 1951.

—. *Enarrationes in Psalmos.* Ed. Eloi Dekkers and Jean Fraipont. 3 vols. CCSL 38–40. Turnhout: Brepols, 1956.

—. *Expositions of the Psalms.* Trans. Maria Boulding. 6 vols. WSA, Part 3, vols. 15–20. Hyde Park, NY: New City Press, 2000–2004.

—. *Expositions on the Psalms.* Trans. J.E. Tweed. NPNF1, 8.

—. *Epistulae.* Ed. Alois Goldbacher. 5 vols. CSEL 34.1, 34.2, 44, 57–58. Vienna and Leipzig: Tempsky, 1895–1923.

—. *Letters.* Trans. Roland J. Teske. 4 vols. WSA, Part 2, vols. 1–4. Hyde Park, NY: New City Press, 2001–2005.

—. *Letters.* Trans. Sister Wilfrid Parsons. 5 vols. FOTC 12, 18, 20, 30, 32. Washington, DC: Catholic University of America Press, 1951–1964.

—. *Letters 1*–29*.* Trans. Robert B. Eno. FOTC 81. Washington, DC: Catholic University of America Press, 1989.

—. *Letters of St. Augustine of Hippo.* Trans. J.G. Cunningham. NPNF1, 1.

—. *In Epistolam Ioannis ad Parthos tractatus x*. PL 35:1977–2062.

—. *Homilies on the First Epistle of John*. Trans. Boniface Ramsey. WSA, Part 3, vol. 14. Hyde Park, NY: New City Press, 2008.

—. *Homilies on the First Epistle of John*. Trans. H. Browne. NPNF1, 7.

—. *De Genesi ad litteram libri duodecim*. Ed. Joseph Zycha. CSEL 28.1. Vienna: F. Tempsky, 1894. Pp. 1–456.

—. *The Literal Meaning of Genesis*. Trans. John Hammond Taylor. 2 vols. Ancient Christian Writers, 41–42. New York: Newman Press, 1982.

—. *De magistro*. Ed. Klaus-Detlef Daur. In *Contra academicos; De beata vita; De ordine*. CCSL 29. Turnhout: Brepols, 1970. Pp. 139–203.

—. *Against the Academicians; The Teacher*. Trans. Peter King. Indianapolis: Hackett, 1995.

—. *Sermones (classis iv)*. PL 38:23–39:1736.

—. *Sermons*. Trans. Edmund Hill. 11 vols. WSA, Part 3, vols. 1–11. Hyde Park, NY: New City Press, 1990–1997.

—. *Soliloquiorum libri duo*. Ed. Wolfgang Hörmann. In *Soliloquiorum libri duo, De inmortalitate animae, De quantitate animae*. CSEL 89. Vienna: Hoelder-Pichler-Tempsky, 1986. Pp. 1–98.

—. *Soliloquies; and, Immortality of the Soul*. Trans. Gerard Watson. Warminster, Engl.: Aris & Phillips, 1990.

—. *Soliloquies*. Trans. C.C. Starbuck. NPNF1, 7.

—. *De trinitate*. Ed. William J. Mountain and François Glorie. 2 vols. CCSL 50, 50A. Turnhout: Brepols, 1968.

—. *The Trinity*. Trans. Edmund Hill. WSA, Part 1, vol. 5. Hyde Park, NY: New City Press, 1991.

—. *The Trinity*. Trans. Stephen McKenna. FOTC 45. Washington, DC: Catholic University of America Press, [1963; reprint 1981].

—. *On the Trinity*. Trans. Arthur West Haddan. NPNF1, 3.

—. *De utilitate credendi*. In *De utilitate credendi. De duabus animabus. Contra Fortunatum. Contra Adimantum. Contra epistulam fundamenti. Contra Faustum*. Ed. Joseph Zycha. CSEL 25.1. Vienna: F. Tempsky, 1891. Reprint New York: Johnson Reprint Corp., 1972. Pp. 3–48.

—. *De utilitate credendi ad Honoratum liber unus*. PL 42:65–92.

—. *The Advantage of Believing*. Trans. Ray Kearney. In *On Christian Belief*. Ed. Boniface Ramsey. WSA, Part 1, vol. 8. Hyde Park, NY: New City Press, 2005. Pp. 116–48.

—. *The Usefulness of Belief*. In *Earlier Writings*. Trans. John H.S. Burleigh. Library of Christian Classics, 6. Philadelphia: Westminster Press, 1953. Pp. 291–323.

—. *On the Profit of Believing*. Trans. C.L. Cornish. NPNF1, 3.

Ps.-Augustine of Hippo. *Sermones*. PL 39.

Avianus. *Fabulae*. Ed. and trans. J. Wight Duff and Arnold M. Duff. In *Minor Latin Poets*. LCL 284. Cambridge, MA: Harvard University Press; London: W. Heinemann, 1934; reprint 1968. Pp. 667–749.

Avicenna. *De animalibus*. In *Opera in lucem redacta ac nuper quantum ars niti potuit per canonicos emendata*. Venice, 1508; reprint Frankfurt/M: Minerva, 1961. Fols. D5ra–H8rb (29ra–64rb).

Bacon, Roger. *De multiplicatione specierum*. Ed. and trans. David C. Lindberg. In *Roger Bacon's Philosophy of Nature. A Critical Edition, with English Translation, Introduction, and Notes, of De multiplicatione specierum and De speculis comburentibus*. Oxford: Clarendon, 1983.

—. *The 'Opus majus' of Roger Bacon*. Ed. John Henry Bridges. 2 vols. Oxford: Clarendon, 1897.

—. *The Opus majus of Roger Bacon*. Trans. Robert Belle Burke. 2 vols. Philadelphia: University of Pennsylvania Press; London: H. Milford, Oxford University Press, 1928; reprint Bristol, Engl., and Sterling, VA: Thoemmes Press, 2000.

—. *Perspectiva*. Ed. and trans. David C. Lindberg. In *Roger Bacon and the Origins of Perspectiva in the Middle Ages. A Critical Edition and English Translation of Bacon's Perspectiva with Introduction and Notes*. Oxford: Clarendon; New York: Oxford University Press, 1996.

Bartholomeus Anglicus. *De proprietatibus rerum*. Nuremberg: A<n>thonius Koburger, 1492.

—. Trans. John Trevisa. Ed. M.C. Seymour et al. In *On the Properties of Things. John Trevisa's Translation of Bartholomaeus Anglicus, De Proprietatibus Rerum. A Critical Text*. 3 vols. Oxford: Clarendon, 1975–1988.

Bede the Venerable. *Historia ecclesiastica gentis Anglorum*. Ed. and trans. Bertram Colgrave and Roger A.B. Mynors. In *Bede's Ecclesiastical History of the English People*. Oxford: Clarendon, 1969.

Bernard of Clairvaux. *Apologia ad Guillelmum abbatem*. Ed. Jean Leclercq and Henri M. Rochais. In *Opera*. 8 vols. Rome: Editiones Cistercienses, 1957–1977. Vol. 3. Pp. 61–108.

—. *Cistercians and Cluniacs. St. Bernard's Apologia to Abbot William*. Trans. Michael Casey. CFS 1. Kalamazoo, MI: Cistercian Publications, 1970.

—. *Epistolae*. Ed. Jean Leclercq and Henri M. Rochais. 2 vols. In *Opera*. 8 vols. Rome: Editiones Cistercienses, 1957–1977. Vols. 7–8.

—. *The Letters of St. Bernard of Clairvaux*. Trans. Bruno Scott James. CFS 62. Kalamazoo, MI: Cistercian Publications, 1998.

—. *Homiliae in laudibus Virginis matris*. Ed. Jean Leclercq and Henri M. Rochais. In *Opera*. 8 vols. Rome: Editiones Cistercienses, 1957–1977. Vol. 4. Pp. 1–58.

—. *Magnificat: Homilies in Praise of the Blessed Virgin Mary. By Bernard of Clairvaux and Amadeus of Lausanne*. Trans. Marie-Bernard Saïd and Grace Perigo. CFS 18. Kalamazoo, MI: Cistercian Publications, 1979.

—. *Liber de gradibus humilitatis et superbiae*. Ed. Jean Leclercq and Henri M. Rochais. In *Opera*. 8 vols. Rome: Editiones Cistercienses, 1957–1977. Vol. 3. Pp. 1–59.

—. *The Steps of Humility*. Trans. George Bosworth Burch. Cambridge, MA: Harvard University Press, 1940.

—. *Sermo in Dominica infra octavam assumptionis*. Ed. Jean Leclercq and Henri M. Rochais. In *Opera*. 8 vols. Rome: Editiones Cistercienses, 1957–1977. Vol. 5. Pp. 262–74.

—. *St. Bernard's Sermons on the Blessed Virgin Mary*. Trans. a Priest of Mount Melleray [Ailbe J. Luddy]. Chumleigh: Augustine Publishing Company, 1984; reprint 1987.

—. *Sermones de diversis*. Ed. Jean Leclercq and Henri M. Rochais. In *Opera*. 8 vols. Rome: Editiones Cistercienses, 1957–1977. Vol. 6/1. Pp. 57–406.

—. *Sermones in Psalmum "Qui habitat."* Ed. Jean Leclercq and Henri M. Rochais. In *Opera*. 8 vols. Rome: Editiones Cistercienses, 1957–1977. Vol. 4. Pp. 381–492.

—. *Sermons on Conversion*. Trans. Marie-Bernard Saïd. CFS 25. Kalamazoo, MI: Cistercian Publications, 1981.

—. *Sermones super Cantica canticorum*. Ed. Jean Leclercq, Charles H. Talbot, and Henri M. Rochais. In *Opera*. 8 vols. Rome: Editiones Cistercienses, 1957–1977. Vols. 1–2.

—. *On the Song of Songs*. Trans. Kilian Walsh and Irene M. Edmonds. 4 vols. CFS 4, 7, 31, 40. Spencer, MA: Cistercian Publications, 1971–1980.

—. *Tractatus de consideratione ad Eugenium papam*. Ed. Jean Leclercq and Henri M. Rochais. In *Opera*. 8 vols. Rome: Editiones Cistercienses, 1957–1977. Vol. 3. Pp. 379–493.

—. *Five Books on Consideration: Advice to a Pope*. Trans. John D. Anderson and Elizabeth T. Kennan. CFS 37. Kalamazoo, MI: Cistercian Publications, 1976.

Ps.-Bernard of Clairvaux. *Meditationes piissimae de cognitione humanae conditionis*. PL 184:485A–508B.

Bernard of Morlaix (Bernard of Cluny) (?). *De vanitate mundi* ("Chartula nostra ..."). PL 184:1307A–1314C.

Bernard Silvestris. *Cosmographia*. Ed. Peter Dronke. Textus Minores, 53. Leiden: Brill, 1978.

—. *The Cosmographia of Bernardus Silvestris*. Trans. Winthrop Wetherbee. New York: Columbia University Press, 1973; reprint 1990.

Bestiarium. Ed. and trans. Willene B. Clark. In *A Medieval Book of Beasts. The Second-Family Bestiary*. Woodbridge: Boydell Press, 2006.

Biblia sacra iuxta Vulgatam Clementinam. Nova editio. Ed. Alberto Colunga and Laurentio Turrado. 4th ed. Madrid: Biblioteca de autores Cristianos, 1965.

—. Trans. the English College at Rheims, 1582, and the English College at Douay, 1609. Rev. ed. Ed. Richard Challoner. Baltimore and New York: John Murphy, 1899; reprint Rockford IL: TAN Books, 1971. Available at http://www.drbo.org/ (accessed August 4, 2011).

Boethius. *Philosophiae consolatio*. Ed. Ludwig Bieler. CCSL 94. Turnhout: Brepols, 1957.

—. *De consolatione Philosophiae*. Ed. Hugh F. Stewart and Edward K. Rand. Trans. S.J. Tester. In *The Theological Tractates. The Consolation of Philosophy*. LCL 74.

Cambridge, MA: Harvard University Press; London: William Heinemann, Ltd., 1973.

Bonaventure of Balneoreggio. *Sermons de diversis*. Ed. Jacques Guy Bougerol. 2 vols. Paris: Les editions Franciscaines, 1993.

Ps.-Caecilius Balbus. *De nugis philosophorum*. Ed. Eduard Woelfflin. In *De nugis philosophorum quae supersunt*. Basel: Schweighauser, 1855.

Cantiones Bohemicae. Ed. Guido Maria Dreves. Analecta hymnica medii aevi, 1. Leipzig: Reisland, 1886.

Cassian, John. *Conlationes xxiiii*. Ed. Michael Petschenig. CSEL 13. Vienna: Geroldi, 1886.

—. *The Conferences*. Trans. Boniface Ramsey. Ancient Christian Writers, 57. New York: Paulist Press, 1997.

—. *De institutis coenobiorum et de octo principalium vitiorum remediis libri xii*. Ed. Michael Petschenig. CSEL 17. Vienna: Tempsky, 1888.

—. *The Institutes*. Trans. Boniface Ramsey. Ancient Christian Writers, 58. New York: Paulist Press, 2000.

Magnus Aurelius Cassiodorus. *Variarum libri xii*. Ed. Åke J. Fridh. CCSL 96. Turnhout: Brepols, 1973.

—. *The Variae of Magnus Aurelius Cassiodorus ... Being Documents of the Kingdom of the Ostrogoths in Italy, Chosen to Illustrate the Life of the Author and the History of His Family*. Trans. S.J.B. Barnish. Translated Texts for Historians, 12. Liverpool: Liverpool University Press, 1992.

—. *The Letters of Cassiodorus, Being a Condensed Translation of the Variae epistolae of Magnus Aurelius Cassiodorus Senator with an Introduction*. Trans. Thomas Hodgkin. London: Frowde, 1886. Available at http://www.gutenberg.org/files/18590/18590-h/18590-h.htm (accessed August 2, 2011).

Cassiodorus-Epiphanius. *Historia ecclesiastica tripartita*. Ed. Walter Jacob and Rudolf Hanslik. CSEL 71. Vienna: Hoelder-Pichler-Tempsky, 1952.

Chartularium Universitatis Parisiensis. Ed. Heinrich Denifle and Emile Chatelain. 4 vols. Paris: Ex typis fratrum Delalain, 1889–1897.

Marcus Tullius Cicero. *De officiis*. Ed. Michael Winterbottom. Oxford: Clarendon, 1994.

—. *On Duties*. Trans. Walter Miller. LCL 21. Cambridge, MA: Harvard University Press, 1913. Available at http://penelope.uchicago.edu/Thayer/E/Roman/Texts/Cicero/de_Officiis/home.html (accessed August 5, 2011).

—. *Oratio de haruspicum responsis*. Ed. William Peterson. In *Orationes*. 6 vols. Oxford: Clarendon, 1901–1911; reprint 1962. Vol. 5. Sig. 7(3)r–9(3)r.

—. *The Speeches*. Trans. Nevile H. Watts. LCL 158. Cambridge, MA: Harvard University Press, 1923; reprint 1935.

—. *Paradoxa*. Ed. Carl F.W. Mueller. In *Scripta quae manserunt omnia*. 10 vols. Leipzig: Teubner, 1878–1905. Pars 4, vol. 3. Pp. 197–213.

—. *On the Orator: Book 3. On Fate. Stoic Paradoxes. Divisions of Oratory*. Trans. Harris

Rackham. Rev. ed. LCL 349. Cambridge, MA: Harvard University Press; London: W. Heinemann, 1948.

—. *De re publica*. Ed. Konrat Ziegler. 7th ed. Leipzig: Teubner, 1969.

—. *On the Republic. On the Laws*. Trans. Clinton W. Keyes. LCL 213. Cambridge, MA: Harvard University Press; London: W. Heinemann, 1928; reprint 1977.

—. *Tusculanae disputationes*. Ed. Max Pohlenz. Stuttgart: Teubner, 1918; reprint 1967.

—. *Tusculan Disputations*. Ed. John E. King. Rev. ed. LCL 141. Cambridge, MA: Harvard University Press, 1945; reprint 1971.

Claudius Claudianus. *In Eutropium*. Ed. John Barrie Hall. In *Carmina*. Leipzig: Teubner, 1985. Pp. 143–89.

—. *In Rufinum*. Ed. John Barrie Hall. In *Carmina*. Leipzig: Teubner, 1985. Pp. 30–51.

—. *Panegyricus dictus Honorio Augusto quartum consuli*. Ed. John Barrie Hall. In *Carmina*. Leipzig: Teubner, 1985. Pp. 61–85.

—. *Claudian*. Trans. Maurice Platnauer. 2 vols. LCL 135–136. Cambridge, MA: Harvard University Press; London: W. Heinemann, 1922; reprint 1956. Available at http://penelope.uchicago.edu/Thayer/E/Roman/Texts/Claudian/home.html (accessed August 5, 2011).

Constantine the African. *Liber de oculis*. Ed. Pierre Pansier. In *Liber de oculis, Galieni littere ad Corisium de morbis oculorum et eorum curis*. Collectio ophtalmologica veterum auctorum, 7. Paris: J.-B. Baillière et fils, 1909–1933. Pp. 161–208.

Defensor of Ligugé (?). *Liber scintillarum*. Ed. Henri M. Rochais. CCSL 117. Turnhout: Brepols, 1957.

Ps.-Dionysius the Areopagite. *De caelesti hierarchia*. Ed. Günter Heil. Corpus Dionysiacum, 2. Patristische Texte und Studien, 36. Berlin and New York: Walter de Gruyter, 1991. Pp. 5–59.

—. *De caelesti hierarchia*. Trans. Robert Grosseteste et al. Ed. Phillipe Chevallier et al. In *Dionysiaca. Recueil donnant l'ensemble des traductions latines des ouvrages attribués au Denys de l'Aréopage* ... . 2 vols. [Bruges:] Desclée de Brouwer, 1937–1950. Vol. 2. Pp. 724–1066.

—. *De divinis nominibus*. Ed. Beate Regina Suchla. Corpus Dionysiacum, 1; Patristische Texte und Studien, 33. Berlin and New York: Walter de Gruyter, 1990.

—. *De divinis nominibus*. Trans. Robert Grosseteste et al. Ed. Phillipe Chevallier et al. In *Dionysiaca: Recueil donnant l'ensemble des traductions latines des ouvrages attribués au Denys de l'Aréopage* ... . 2 vols. [Bruges]: Desclée de Brouwer, 1937. Vol. 1. Pp. 3–561.

—. *The Complete Works*. Trans. Colm Luibheid. New York and Mahwah, NJ: Paulist Press, 1987.

Disticha Catonis. Ed. Marcus Boas. 2nd ed. Ed. Hendrik Johan Botschuyver. Amsterdam: North-Holland Pub. Co., 1952.

—. *The Distichs of Cato. A Famous Medieval Textbook*. Trans. Wayland Johnson Chase. University of Wisconsin Studies in the Social Sciences and History, 7. Madison: n.p., 1922. Available through Google Books.

Aelius Donatus. *Ars Donati Grammatici urbis Romae – Editio prima (Ars minor)*. Ed. Louis Holtz. In *Donat et la tradition de l'enseignement grammatical. Étude sur l'Ars Donati et sa diffusion (IVe – IXe siècle) et édition critique*. Documents, Études et Répertoires publiés par l'Institut de Recherche et d'Histoire des Textes, 24. Paris: Centre Nationale de la Recherche Scientifique, 1981.

—. *The Ars minor of Donatus, for One Thousand Years the Leading Textbook of Grammar*. Trans. Wayland Johnson Chase. University of Wisconsin studies in the social sciences and history, 11. Madison: n.p., 1926.

Durand, William, of Mende. *Rationale divinorum officiorum*. Ed. Anselme Davril and Timothy M. Thibodeau. 3 vols. CCCM 140, 140A, 140B. Turnhout: Brepols, 1995–2000.

—. *The Rationale divinorum officiorum of William Durand of Mende: A New Translation of the Prologue and Book One*. Trans. Timothy M. Thibodeau. New York: Columbia University Press, 2007.

—. *William Durand on the Clergy and Their Vestments, A New Translation of Books 2 and 3 of The Rationale divinorum officiorum*. Trans. Timothy M. Thibodeau. [Scranton, PA]: University of Scranton Press, 2010.

"An Edition of Five Medieval Sermons from MS Laud misc. 706." Ed. Patrick J. Horner. Ph.D. Dissertation, State University of New York at Albany, 1975.

Magnus Felix Ennodius. *Opera*. Ed. Friedrich Vogel. MGH. Auctores antiquissimi, 7. Berlin: Weidmann, 1885.

Epistola Alexandri regis ad Dindimum regem. Ed. Friedrich Pfister. In *Kleine Texte zum Alexanderroman*. Sammlung vulgärlateinischer Texte, 4. Heidelberg: Carl Winter, 1910. P. 10.

Epistola Dindimi prima ad Alexandrum. Ed. Pfister. In *Kleine Texte zum Alexanderroman*. Pp. 11–16.

Epitome vitae sancti Hugonis abbatis Cluniacensis ab Ezelone atque Gilone monachis Cluniacensibus proxime ab obitu Sancti scriptæ, per Anonymum excerpta. In *Acta sanctorum Aprilis collecta, digesta, illustrata*. Ed. Godefroid Henschen and Daniel van Papenbroeck. 3 vols. Antwerp: Michael Cnobarus, 1675. Vol. 3 (29 April). Cols. 655E–58B.

Euclid. *De speculis*. Ed. Ken'ichi Takahashi. In *The Medieval Latin Traditions of Euclid's Catoptrica. A Critical Edition of De speculis with an Introduction, English Translation and Commentary*. Fukuoka-shi: Kyushu University Press, 1992. Pp. 113–209.

—. *Liber de visu*. Ed. Wilfred R. Theisen. "*Liber de visu*: The Greco-Latin Translation of Euclid's *Optics*." *Mediaeval Studies* 41 (1979): 44–105.

Eusebius of Caesarea. *Chronicle*. Trans. Jerome (Saint). Ed. Rudolf Helm. In *Eusebius Werke 7: Die Chronik des Hieronymus*. Die griechischen christlichen Schriftsteller der ersten drei Jahrhunderte, 47. 2nd ed. Leipzig: Hinrichs, 1956; reprint Berlin: Akademie-Verlag, 1984.

—. *Chronicle*. Trans. Jerome (Saint). In *The Chronicle of St. Jerome*. Trans. Roger Pearse et al. Available at http://www.tertullian.org/fathers/jerome_chronicle_00_eintro.htm (accessed August 3, 2011).

[*Exempla*]. Frederic C. Tubach. *Index exemplorum. A Handbook of Medieval Religious Tales*. FF Communications, 204. Helsinki: Academia Scientiarum Fennica, 1969; 2nd ed. 1981.

Flos medicinae Scholae Salerni. Ed. Salvatore de Renzi. In *Collectio Salernitana: Ossia documenti inediti, e trattati di medicina appartenenti alla scuola medica Salernitana, raccolti ed illustrati da G.E.T. Henschel, C. Daremberg, e S. de Renzi*. 5 vols. Naples: Dalla tipografia del Filiatre-Sebezio, 1852–1859; reprint Naples: D'Auria Editore, 2001. Vol. 1. Pp. 445–516.

Venantius Fortunatus. *Carmina*. Ed. Friedrich Leo. In *Opera poetica*. MGH, Auctores antiquissimi, 4.1. Berlin: Weidmann, 1881.

—. *Poèmes*. Ed. and French trans. Marc Reydellet. 3 vols. Paris: Les belles lettres, 1994–2004.

Sextus Julius Frontinus. *Strategemata*. Ed. Robert I. Ireland. Leipzig: Teubner, 1990.

—. *Stratagems. Aqueducts of Rome*. Trans. Charles E. Bennett and Clemens Herschel. LCL 174. Cambridge, MA: Harvard University Press, 1925; reprint 1980. Available at http://penelope.uchicago.edu/Thayer/E/Roman/Texts/Frontinus/Strategemata/1*.html#1 (accessed August 3, 2011).

Fulgentius the Mythographer. *Mitologiarum libri tres*. Ed. Rudolf Helm and Jean Préaux. In *Opera*. Stuttgart: Teubner, 1970.

—. *Fulgentius the Mythographer*. Trans. Leslie George Whitbread. [Columbus:] Ohio State University Press, 1971.

Aulus Gellius. *Noctes Atticae*. Ed. Peter K. Marshall. 2 vols. Oxford: Clarendon, 1990.

—. *Attic Nights*. Trans. John C. Rolfe. 3 vols. LCL 195, 200, 212. Cambridge, MA: Harvard University Press; London: W. Heinemann, 1967–1984. Books 1–13 available at http://penelope.uchicago.edu/Thayer/E/Roman/Texts/Gellius/home.html (accessed August 3, 2011).

Geoffrey of Vinsauf. *Poetria nova*. Ed. Edmond Faral. In *Les arts poétiques du XIIe et du XIIIe siècle. Recherches et documents sur la technique littéraire du moyen age*. Paris: Champion, 1962. Pp. 194–262.

—. *Poetria nova*. Trans. Margaret F. Nims. Rev. ed. MST 49. Toronto: Pontifical Institute of Mediaeval Studies, 2010.

Gerard of Cremona (?). *Praefatio in Almagestum*. Ed. and trans. Charles Burnett. "'Ptolemaeus in Almagesto dixit': The Transformation of Ptolemy's *Almagest* in Its Transmission via Arabic into Latin." In *Transformationen antiker Wissenschaften*. Ed. Georg Toepfer and Hartmut Böhme. Berlin and New York: Walter de Gruyter, 2010. Pp. 115–40, here 125–30.

— (?). *Praefatio*. In *Almagestum Cl. Ptolemei Pheludiensis Alexandrini Astronomorum principis: Opus ingens ac nobile omnes Celorum motus continens. Felicibus Astris eat in lucem: Ductu Petri Liechtenstein Coloniensis Germani*. Venice: ex officina eiusdem litteraria, 1515.

Gesta Romanorum. Ed. Hermann Oesterley. Berlin: Weidmann, 1872.

—. Trans. Charles Swan. Rev. and corr. Wynnard Hooper. London: George Bell & Sons, 1906. Available through Google Books.

Gilbert of Tournai. *Collectio de scandalis ecclesiae*. Ed. Autbertus Stroick. In "Collectio de scandalis ecclesiae. Nova Editio." *Archivum Franciscanum Historicum* 24 (1931): 33–62.

Glossa ordinaria. In *Biblia latina cum Glossa ordinaria. Facsimile Reprint of the Editio Princeps, Adolph Rusch of Strassburg 1480/81*. 4 vols. Turnhout: Brepols, 1992.

—. 2 vols. PL 113:11B–114:752B.

Gratian. *Decretum Magistri Gratiani*. Ed. Emil Ludwig Richter. 2nd ed. Ed. Emil Friedberg. In *Corpus iuris canonici*. 2 vols. Leipzig: Tauchnitz, 1879–1881. Vol. 1.

Gregory the Great. *Dialogi*. In *Dialogues*. Trans. Paul Antin. Ed. Adalbert de Vogüé. 3 vols. SC 251, 260, 265. Paris: Cerf, 1978–1980.

—. *The Dialogues of Saint Gregory, surnamed the Great: Pope of Rome & the first of that name. Divided into Four Books, wherein he entreateth of the Lives and Miracles of the Saints in Italy and of the Eternity of Men's Souls*. Trans. P.W. (1608). Ed. Edmund G. Gardner. London: Philip Lee Warner, 1911. Available at http://www.ccel.org/ccel/pearse/morefathers/files/index.htm#Gregory_Dialogues (accessed August 3, 2011).

—. *Expositio in septem psalmos poenitentiales*. PL 79:549D–658B.

—. *Homiliae in Evangelia*. Ed. Raymond Étaix. CCSL 141. Turnhout: Brepols, 1999.

—. *Forty Gospel Homilies*. Trans. David Hurst. Cistercian Studies Series, 123. Kalamazoo, MI: Cistercian Publications, 1990.

—. *Moralia in Iob*. Ed. Marc Adriaen. 3 vols. CCSL 143, 143A, 143B. Turnhout: Brepols, 1979–1985.

—. *Morals on the Book of Job*. Trans. with Notes and Indices. 3 vols. Oxford: J.H. Parker, 1844–1850. Available at http://www.lectionarycentral.com/GregoryMoraliaIndex.html (accessed August 3, 2011).

—. *Registrum epistularum libri xiv*. Ed. Dag Norberg. 2 vols. CCSL 140, 140A. Turnhout: Brepols, 1982.

—. *The Letters of Gregory the Great*. Trans. John R.C. Martyn. 3 vols. MST 40. Toronto: Pontifical Institute of Mediaeval Studies, 2004.

—. *Regulae pastoralis liber*. PL 77:13A–128A.

—. *Pastoral Rule*. Trans. James Barmby. NPNF2, 12.

Gregory of Tours. *Libri historiarum decem*. Ed. Bruno Krusch and Wilhelm Levison. MGH, Scriptores rerum Merovingicarum, 1/1. 2nd ed. Hannover: Hahn, 1951.

—. *The History of the Franks*. Trans. Lewis Thorpe. Harmondsworth and Baltimore: Penguin, 1974.

Helinand of Froidmont. *Chronicon*. PL 212:771C–1082C.

—. *Flores a Vincentio Bellovacensi collecti*. PL 212:721A–760A.

—. *Sermones*. PL 212:481C–720C.

Hermannus de Runa. *Sermones festivales*. Ed. Edmund Mikkers, Joseph Theuws, and Roland Demeulenaere. CCCM 64. Turnhout: Brepols, 1986.

Holcot, Robert. *Liber moralizationum historiarum*. In *In Librum Sapientiae Regis Salomonis Praelectiones CCXIII ...* . Basel: n.p., 1586. Pp. 708–50.

Honorius of Autun. *Speculum ecclesiae*. PL 172:807A–1108A.

Horace. *De arte poetica*. Ed. Edward C. Wickham. 2nd ed. Ed. Heathcote W. Garrod. In *Opera*. Oxford: Clarendon, 1901.

—. *Epistulae*. Ed. Edward C. Wickham. 2nd ed. Ed. Heathcote W. Garrod. In *Opera*. Oxford: Clarendon, 1901.

—. *Sermones*. Ed. Edward C. Wickham. 2nd ed. Ed. Heathcote W. Garrod. In *Opera*. Oxford: Clarendon, 1901.

—. *Satires, Epistles and Ars Poetica*. Trans. Henry Rushton Fairclough. Rev. ed. LCL 194. Cambridge, MA: Harvard University Press, 1929; reprint 1966.

Hrabanus Maurus. *Expositio in Matthaeum*. Ed. Bengt Löfstedt. 2 vols. CCCM 174, 174A. Turnhout: Brepols, 2000.

Hugh of Fouilloy (Ps.-Hugh of Saint-Victor). *Aviarium*. Ed. and trans. Willene B. Clark. In *The Medieval Book of Birds. Hugh of Fouilloy's Aviarium*. Medieval & Renaissance Texts & Studies, 80. Binghamton, NY: Medieval & Renaissance Texts & Studies, 1992.

Hugh of Saint-Cher (?). *Expositio super Apocalypsim*. In Thomas Aquinas. *Opera omnia*. 25 vols. Parma: Tipis Petri Fiaccadori, 1852–1873. Available at http://www.corpusthomisticum.org/x1a00.html (accessed December 10, 2010).

Hugh of Saint-Victor. *De claustro anime libri iv*. PL 176:1017B–1184A.

—. *Commentariorum in Hierarchiam Coelestem s. Dionysii Areopagitae libri x*. PL 175:923B–1154C.

—. *Didascalicon*. PL 176:739C–838D.

—. *Didascalicon*. Ed. Charles Henry Buttimer. The Catholic University of America Studies in Medieval and Renaissance Latin, 10. Washington, DC: The Catholic University Press, 1939.

—. *The Didascalicon of Hugh of St. Victor: A Medieval Guide to the Arts*. Trans. Jerome Taylor. New York: Columbia University Press, 1961; reprint 1991.

—. *In Salomonis Ecclesiasten Homiliae xix*. PL 175:113A–256C.

—. *Liber de institutione novitiorum*. PL 176:925A–62B.

—. *De la Formation des Novices*. In *L'œuvre de Hugues de Saint-Victor*. Ed. Hugh B.

Feiss and Patrice Sicard. French trans. Dominique Poirel, Henri Rochais, and Patrice Sicard. Vol. 1. Sous la règle de Saint Augustin, 3. Turnhout: Brepols, 1997.

—. *De sacramentis Christianae fidei*. PL 176:173A–618B.

—. *On the Sacraments of the Christian Faith*. Trans. Roy Deferrari. Publications of the Mediaeval Academy of America, 58. Cambridge, MA: Mediaeval Academy of America, 1951.

Ps.-Hugh of St.-Victor. *De bestiis et aliis rebus. Libri iii–iv*. PL 177:83A–164D.

Iacopo da Varazze (Jacobus de Voragine). *Chronica civitatis Ianuensis*. Ed. Giovanni Monleone. In *Iacopo da Varagine e la sua Cronaca di Genova dalle origini al MCCXCVII*. 3 vols. Fonti per la storia d'Italia pubblicate dal R. Istituto Storico Italiano per il Medio Evo. Scrittori, secolo XIII, [84–86]. Rome: Tipografia del Senato, 1941.

—. *Legenda aurea. Con le miniature del codice Ambrosiano C 240 inf.* Ed. Giovanni Paolo Maggioni. Italian trans. Gianfranco Agosti et al. 2 vols. Edizione nazionale dei testi mediolatini, 20; Serie II, 9. Florence: SISMEL edizioni del Galluzzo; Milan: Biblioteca Ambrosiana, 2007.

—. *The Golden Legend: Readings on the Saints*. Trans. William Granger Ryan. 2 vols. Princeton: Princeton University Press, 1993.

In Matthaeum. Ed. Bengt Löfstedt. CCCM 159. Turnhout: Brepols, 2003.

Innocent III, Pope. *De miseria conditionis humane*. Ed. and trans. Robert E. Lewis. Athens, GA: The University of Georgia Press, 1978.

Isidore of Seville. *Allegoriae quaedam Sacrae Scripturae*. PL 83:97B–130B.

—. *Etymologiarum siue originum libri xx*. Ed. Wallace M. Lindsay. 2 vols. Oxford: Clarendon, 1911.

—. *The Etymologies of Isidore of Seville*. Trans. Stephen A. Barney et al. Cambridge, Engl., and New York: Cambridge University Press, 2006.

—. *Sententiae*. Ed. Pierre Cazier. CCSL 111. Turnhout: Brepols, 1998.

—. *Synonyma*. Ed. Jacques Elfassi. CCSL 111B. Turnhout: Brepols, 2009.

Jerome (Saint). *Aduersus Iouinianum libri duo*. PL 23:211A–338B.

—. *Against Jovinianus*. Trans. W.H. Fremantle, G. Lewis, and W.G. Martley. NPNF2, 6.

—. *Epistula aduersus Rufinum*. Ed. Pierre Lardet. CCSL 79. Turnhout: Brepols, 1982. Pp. 73–116.

—. *The Apology against the Books of Rufinus*. Trans. John Nicholas Hritzu. In *Dogmatic and Polemical Works*. FOTC 53. Washington, DC: Catholic University of America Press, 1965. Pp. 47–220.

—. *Apology Against Rufinus (Book III)*. Trans. W.H. Fremantle. NPNF2, 3.

—. *Epistulae*. Ed. Isidor Hilberg. 3 vols. CSEL 54–56. Vienna: F. Tempsky; Leipzig: Freytag, 1910–1918.

—. *Letters*. Trans. W.H. Fremantle, G. Lewis, and W.G. Martley. NPNF2, 6.

—. *Liber interpretationis Hebraicorum nominum*. Ed. Paul de Lagarde. CCSL 72. Turnhout: Brepols, 1959. Pp. 57–161.

John Chrysostom. *Commentarius in sanctum Joannem Apostolum et Evangelistam*. PG 59:23–482.

—. *Commentary on Saint John the Apostle and Evangelist*. Trans. Sister Thomas Aquinas Goggin. 2 vols. FOTC 33, 41. Washington, DC: Catholic University of America Press, [1957–1959; reprint 1969–1977].

—. *In Matthaeum homiliae xc*. 2 vols. PG 57:13–58:794.

—. *Homilies on Matthew*. Trans. George Prevost. Rev. M.B. Riddle. NPNF1, 10.

Ps.-John Chrysostom. *Opus imperfectum in Matthaeum*. PG 56:611–946.

—. *Incomplete Commentary on Matthew (Opus imperfectum)*. Trans. James A. Kellerman. Ed. Thomas C. Oden. Downers Grove, IL: InterVarsity Press, 2010.

Johannes de Hauvilla. *Architrenius*. Ed. Paul Gerhard Schmidt. Munich: Fink, 1974.

—. *Architrenius*. Ed. and trans. Winthrop Wetherbee. Cambridge Medieval Classics, 3. Cambridge, Engl.: Cambridge University Press, 1994.

John of Salisbury. *Policraticus*. Ed. Clemens C.I. Webb. In *Policratici sive De nugis curialium et vestigiis philosophorum libri viii*. 2 vols. Oxford: Clarendon, 1909; reprint New York: Arno Press, 1979.

—. *Policraticus I–IV*. Ed. Katharine S.B. Keats-Rohan. CCCM 118. Turnhout: Brepols, 1993.

—. *Policraticus. Of the Frivolities of Courtiers and the Footprints of Philosophers*. Ed. and trans. Cary J. Nederman. Cambridge, Engl.: Cambridge University Press, 1990.

Flavius Josephus. *Jewish Antiquities*. Ed. and trans. Henry St.J. Thackeray, Ralph Marcus, Allen Wikgren, and Louis H. Feldman. 6 vols. In *Josephus*, 4–9. LCL 242, 281, 326, 365, 410, 433. Cambridge, MA: Harvard University Press; London: William Heinemann, 1930–1965.

—. *The Jewish War*. Ed. and trans. Henry St.J. Thackeray. 2 vols. In *Josephus*, 2–3. LCL 203, 210. Cambridge, MA: Harvard University Press; London: William Heinemann, 1927–1928; reprint 1956, 1961.

Justin (Marcus Junianus Justinus). *Epitoma historiarum Philippicarum Pompei Trogi*. 2nd ed. Ed. Otto Seel. Stuttgart: Teubner, 1985.

—. *Epitome of the Philippic History of Pompeius Trogus*. Trans. John Selby Watson. London: Henry G. Bohn, 1853. Available at http://www.forumromanum.org/literature/justin/english/trans1.html (accessed August 4, 2011).

Juvenal. *Saturae*. Ed. Wendell V. Clausen. In *A. Persi Flacci et D. Juni Juvenalis saturae*. Rev. ed. Oxford: Clarendon, 1992.

—. *Juvenal and Persius*. Ed. and trans. Susanna Morton Braund. LCL 91. Cambridge, MA: Harvard University Press, 2004.

Leontius of Naples. *Vita s. Ioannis Eleemosynarii*. Trans. Anastasius Bibliothecarius. PL 73:337D–84B.

—. *St. John the Almoner*. In *The Lives of the Fathers of the Eastern Deserts: Or the Wonders of God in the Wilderness*. Trans. [from the Greek text] Richard Challoner. New York: D. & J. Sadlier & Co., 1868.

Liber Pharetrae. In Bonaventure of Balneoreggio. *Opera omnia*. Ed. Adolphe C. Peltier. 15 vols. Paris: Vivès, 1864–1871. Vol. 7. Pp. 3–231.

Macrobius. *Saturnalia*. Ed. James Willis. Leipzig: Teubner, 1963.

—. *Saturnalia*. Ed. and trans. Robert A. Kaster. 3 vols. LCL 510–12. Cambridge, MA: Harvard University Press, 2011.

Manegold of Lautenbach (?) (Ps.-Bede the Venerable). *Exegesis de Psalmorum libro*. PL 93:477B–1098C.

Martial. *Epigrammata*. Ed. Wallace M. Lindsay. 2nd ed. Oxford: Clarendon, 1929.

—. *Epigrams*. Ed. and trans. David R. Shackleton Bailey. LCL 94–95, 480. Cambridge, MA: Harvard University Press, 1993.

The New English Bible with the Apocrypha. New York: Oxford University Press, 1971.

Odo of Cheriton. *Parabolae*. Ed. Léopold Hervieux. In *Les fabulistes latins depuis le siècle d'Auguste jusqu'à la fin du moyen âge*. 2nd ed. 5 vols. Paris: Firmin-Didot et Cie., 1893–1899; reprint New York: Burt Franklin, [1965]. Vol. 4. Pp. 265–343.

Odonis de Ceritona fabulis addita. Ed. Léopold Hervieux. In *Les fabulistes latins depuis le siècle d'Auguste jusqu'à la fin du moyen âge*. 2nd ed. 5 vols. Paris: Firmin-Didot et Cie., 1893–1899; reprint New York: Burt Franklin, [1965]. Vol. 4. Pp. 361–416.

Odo of Cluny. *Matins Antiphons for the Feast of Saint Martin*. In *Historia Sancti Martini*. Ed. Martha Fickett. Wissenschaftliche Abhandlungen / Musicological Studies, 65/11. Ottawa: The Institute of Mediaeval Music, 2006. Pp. 5–7, 12–14.

Paulus Orosius. *Historiarum adversum paganos libri vii*. Ed. Karl Zangemeister. CSEL 5. Vienna: Tempsky, 1882.

—. *Seven Books of History against the Pagans*. Trans. A.T. Fear. Translated Texts for Historians, 54. Liverpool: Liverpool University Press, 2010.

Ovid. *Amores*. Ed. Edwin J. Kenney. In *Amores, Medicamina faciei femineae, Ars amatoria, Remedia amoris*. 2nd ed. Oxford: Clarendon, 1994. Pp. 1–108.

—. *Heroides. Amores*. Trans. Grant Showerman. 2nd rev. ed. Ed. George Patrick Goold. LCL 41. Cambridge, MA: Harvard University Press; London: W. Heinemann, 1977.

—. *Ars amatoria*. Ed. Edwin J. Kenney. In *Amores, Medicamina faciei femineae, Ars amatoria, Remedia amoris*. 2nd ed. Oxford: Clarendon, 1994. Pp. 117–219.

—. *The Art of Love and Other Poems*. Trans. John H. Mozley. 2nd rev. ed. Ed. George Patrick Goold. LCL 232. Cambridge, MA: Harvard University Press; London: W. Heinemann, 1979.

—. *Ex Ponto libri quattuor*. Ed. John A. Richmond. Leipzig: Teubner, 1990.

—. *Tristia. Ex Ponto*. Trans. Arthur L. Wheeler. 2nd rev. ed. Ed. George Patrick Goold. LCL 151. Cambridge, MA: Harvard University Press; London: W. Heinemann, 1988.

—. *Metamorphoses*. Ed. William S. Anderson. Leipzig: Teubner, 1977.

—. *Metamorphoses*. Trans. Frank Justus Miller. 3rd rev. ed. Ed. George Patrick Goold. 2 vols. LCL 42–43. Cambridge, MA: Harvard University Press; London: W. Heinemann, 1977–1984.

—. *Remedia amoris*. Ed. Edwin J. Kenney. In *Amores, Medicamina faciei femineae, Ars amatoria, Remedia amoris*. 2nd ed. Oxford: Clarendon, 1994. Pp. 221–61.

—. *Remedies for Love*. In *The Art of Love and Other Poems*. Trans. John H. Mozley. 2nd rev. ed. Ed. George Patrick Goold. LCL 232. Cambridge, MA: Harvard University Press; London: W. Heinemann, 1979.

Paul the Deacon. *Historia Romana. Libri XI–XVI*. Ed. Hans Droysen. In *Eutropi Breviarium ab urbe condita cum versionibus Graecis et Pauli Landolfique additamentis*. MGH. Auctores antiquissimi, 2. Berlin: Weidmann, 1879. Pp. 183–224. Reprint as *Pauli Historia Romana*. MGH. Scriptores rerum Germanicarum in usum scholarum separatim editi, 49. Munich: Monumenta Germaniae Historica, 1978.

—. *Vita beatissimi Gregorii papae urbis Romae*. Ed. Hartmann Grisar. In "Die Gregorbiographie des Paulus Diakonus in ihrer ursprünglichen Gestalt, nach italienischen Handschriften." *Zeitschrift für katholische Theologie* 11 (1887): 158–72.

Pecham, John. *Perspectiva communis*. Ed. and trans. David C. Lindberg. In *John Pecham and the Science of Optics: Perspectiva communis*. Madison, etc.: The University of Wisconsin Press, 1970.

Peraldus, William. *De eruditione principum*. In Thomas Aquinas. *Opera omnia*. 25 vols. Parma: Tipis Petri Fiaccadori, 1852–1873. Available at http://www.corpusthomisticum.org/xreo.html (accessed December 10, 2010).

—. *Summa de vitiis*. In *Summa virtutum ac vitiorum Guilhelmi Paraldi Episcopi Lugdunensis de Ordine Predicatorum*. 2 vols. Paris: Johannes Petit, Johannes Frellon, Franciscus Regnault, 1512. Vol. 2.

Petronius. *Satyricon*. Ed. Konrad Mueller. In *Satyricon reliquiae*. 4th ed. Leipzig: Teubner, 1995.

—. *Satyricon*. Trans. Michael Heseltine. Rev. ed. Ed. Eric H. Warmington. LCL 15. Cambridge, MA: Harvard University Press; London: W. Heinemann, 1913; reprint 1975.

Petrus Alfonsi. *Disciplina clericalis*. Ed. Alfons Hilka and Werner Söderhjelm. Acta Societatis Scientiarum Fennicae, 38.4. Pars 1. Helsingfors: Drukerei der Finnischen litteraturgesellschaft, 1911.

—. *The Disciplina clericalis of Petrus Alfonsi: Translated and Edited by Eberhard Hermes*. Trans. P.R. Quarrie. London: Routledge and Kegan Paul, 1977.

Peter of Blois. *Epistulae*. PL 207:1A–560C.

Peter the Chanter (Petrus Cantor). *Verbum abbreviatum* (textus conflatus). Ed. Monique Boutry. CCCM 196. Turnhout: Brepols, 2004.

Peter Comestor. *Scholastica historia*. PL 198:1049A–722A.

—. *Scholastica historia: Liber Genesis*. Ed. Agneta Sylwan. CCCM 191. Turnhout: Brepols, 2005.

Peter Damiani. *Epistolae*. Ed. Kurt Reindel. In *Die Briefe des Petrus Damiani*. 4 vols. MGH. Die Briefe der deutschen Kaiserzeit, 4/1–4. Munich: Monumenta Germaniae Historica, 1983–1993.

—. *Letters*. Trans. Owen J. Blum et al. 6 vols. FOTC. Medieval Continuation, 1–3, 5–7. Washington, DC: Catholic University of America Press, 1992.

—. *Opuscula varia*. PL 145:19C–858A.

Petrus Helias. *Summa super Priscianum*. Ed. Leo Reilly. 2 vols. Studies and Texts, 113. Toronto: Pontifical Institute of Mediaeval Studies, 1993.

Peter of Limoges. *Tractatus moralis de oculo*. In *Johannis Pithsani archiepiscopi Canthuariensis ordinis fratrum minorum liber de oculo morali foeliciter incipit*. Augsburg: A. Sorg, 1475 (?).

—. *Libro de l'occhio morale et spirituale uulgare. Epistola interpretis Theophili Romani, O. Erem. S. Aug., ad F. Dominicum da Ponzo tanquam auctorem, qui vero in breui praeloquio dicit, se haec invenisse in un Libretto. Sub finem verus auctor dicitur reverendo professor di sacra theologia Maestro P. Lacepiera*. Trans. Teofilo Romano. Venice: Joannes Rubeus Vercellensis, 1496.

Peter Lombard. *In totum Psalterium commentarii*. PL 191:35A–1296D.

—. *Sententiae in iv libris distinctae*. Ed. PP. Collegii s. Bonaventurae. 3rd ed. 2 vols. Spicilegium Bonaventurianum, 4–5. Grottaferrata: Editiones Collegii s. Bonaventurae ad Claras Aquas, 1971–1981.

—. *The Sentences*. Trans. Giulio Silano. 4 vols. MST 42–43, 45, 48. Toronto: Pontifical Institute of Mediaeval Studies, 2007–2010.

Physiologus. Ed. Friedrich Maurer. In *Der altdeutsche Physiologus: Die Millstätter Reimfassung und die Wiener Prosa (nebst dem lateinischen Text und dem althochdeutschen Physiologus)*. Altdeutsche Textbibliothek, 67. Tübingen: Niemeyer, 1967. Pp. 75–95.

—. Trans. Michael J. Curley [from the text established by Francis Carmody]. Austin: University of Texas Press, 1979.

Pliny the Younger. *Naturalis historiae libri xxxvii*. Ed. Ludwig von Jan and Karl Mayhoff. 5 vols. Leipzig: Teubner, 1865–1898; reprint 1967.

—. *Natural History*. Trans. Harris Rackham, William H.S. Jones, and D.E. Eichholz. 10 vols. LCL 330, 352, 353, 370, 371, 392, 393, 418, 394, 419. Cambridge, MA: Harvard University Press, 1967–1975.

[*Proverbs*]. Bartlett Jere Whiting and Helen Westcott Whiting. *Proverbs, Sentences, and Proverbial Phrases from English Writings Mainly before 1500*. Cambridge, MA: Harvard University Press, 1968.

Ptolemy. *Almagestum seu magnae constructionis mathematicae opus plane divinum Latina donatum lingua ab Georgio Trapezuntio ... per Lucam Gauricum ... recognitum*. Venice: L. Giunta, 1528.

—. *Ptolemy's Almagest*. Trans. Gerald J. Toomer. London: Duckworth, 1984.

—. *Optica*. Trans. Eugenius Siculus. Ed. and trans. Albert Lejeune. In *L'Optique de Claude Ptolémée dans la version latine d'après l'arabe de l'émir Eugène de Sicile. Édition critique et exégétique augmentée d'une traduction française et de compléments*. 2nd ed. Collection de travaux de l'Académie Internationale d'Histoire des Sciences, 31. Leiden: Brill, 1989.

—. *Ptolemy's Theory of Visual Perception: An English Translation of the Optics With Introduction and Commentary*. Trans. A. Mark Smith. Transactions of the American Philosophical Society, 86/2. Philadelphia: American Philosophical Society, 1996.

Ps.-Ptolemy. See Gerard of Cremona (?).

Publilius Syrus. *Sententiae*. Ed. Otto Friedrich. Berlin: T. Grieben, 1880; reprint Hildesheim: Georg Olms, 1964.

—. *Sententiae*. Available online at http://www.thelatinlibrary.com/syrus.html (accessed December 10, 2010).

—. *The Moral Sayings of Publius* [!] *Syrus, A Roman Slave. From the Latin*. Trans. Darius Lyman. Cleveland, OH: L.E. Barnard & Co., 1856. Available through Google Books.

Quintilian. *Institutionis oratoriae libri xii*. Ed. Ludwig Radermacher. Rev. ed. Ed. Vinzenz Buchheit. 2 vols. Leipzig: Teubner, 1965.

—. *The Orator's Education*. Ed. and trans. Donald A. Russell. 5 vols. LCL 124–27, 494. Cambridge, MA: Harvard University Press, 2001.

Ps.-Quintilian. *Declamationes XIX maiores Quintiliano falso ascriptae*. Ed. Lennart Håkanson. Stuttgart: Teubner, 1982.

Richard of Saint-Victor. *Expositio moralis in Abdiam*. PL 175:371C–406A.

Rufinus of Aquileia. *Praefatio in Gregorii Nazianzeni orationes*. Ed. August Engelbrecht. In *Opera*. CCSL 20. Turnhout: Brepols, 1961. Pp. 255–56.

Sedulius Scotus. *Collectaneum in Apostolum*. Ed. Hermann Josef Frede and Herbert Stanjek. 2 vols. Vetus Latina. Aus der Geschichte der lateinischen Bibel, 31–32. Freiburg i. Br.: Herder, 1996–1997.

Seneca the Elder (L. Annaeus Seneca Maior). *Controversiae*. Ed. and trans. Michael Winterbottom. In *Declamations*. 2 vols. LCL 463–464. Cambridge, MA: Harvard University Press, 1974.

Seneca the Younger (L. Annaeus Seneca). *Ad Lucilium epistolae morales*. Ed. Leighton D. Reynolds. 2 vols. Oxford: Clarendon, 1965; reprint 1987–1989.

—. *Epistulae morales*. Trans. Richard M. Gummere. 3 vols. LCL 75–77. Cambridge, MA: Harvard University Press, 1917–1925; reprint 1979.

—. *De beneficiis libri vii*. Ed. Karl Hosius. In *De beneficiis libri vii, De clementia libri ii*. 2nd ed. Leipzig: Teubner, 1914. Pp. 1–209.

—. *De beneficiis*. In *Moral Essays*. Trans. John W. Basore. Vol. 3. LCL 310. Cambridge, MA: Harvard University Press, 1935; reprint 2001.

—. *De ira*. Ed. Leighton D. Reynolds. In *Dialogorum libri duodecim*. Oxford: Clarendon, 1977. Pp. 39–128.

—. *De ira*. In *Moral Essays*. Trans. John W. Basore. Vol. 1. LCL 214. Cambridge, MA: Harvard University Press, 1928; reprint 1998.

—. *Naturalium quaestionum libri*. Ed. Harry M. Hine. Stuttgart: Teubner, 1996.

—. *Natural Questions*. Trans. Thomas H. Corcoran. 2 vols. LCL 450, 457. Cambridge, MA: Harvard University Press, 1971–1972.

Ps.-Seneca. *Ad Gallionem de remediis fortuitorum liber*. Ed. Friedrich Haase. In *Opera quae supersunt*. 3 vols. Leipzig: Teubner, 1877–1886. Vol. 3. Pp. 446–57.

—. *Lucii Annei Senecae ad Gallioneni* [!] *de remedis fortuitorum ... Lately translated out of Latyn into Englyshe by Robert Whyttynton* (1547). In Ralph Graham Palmer. *Seneca's De remediis fortuitorum and the Elizabethans*. Chicago: Institute of Elizabethan Studies, 1953.

—. *De moribus liber*. Ed. Friedrich Haase. In *Opera quae supersunt*. 3 vols. Leipzig: Teubner, 1852–1853. Vol. 3. Pp. 462–67. Reprint in *Opera quae supersunt. Supplementum*. Leipzig: Teubner, 1902. Pp. 60–65.

[*Sententiae communes*]. Ed. Hans Walther. *Initia carminum ac versuum medii aevi posterioris latinorum*. Carmina medii aevi posterioris latina, 1. Göttingen: Vandenhoeck & Ruprecht, 1959.

—. Ed. Hans Walther. *Proverbia sententiaeque latinitatis medii aevi*. Carmina medii aevi posterioris latina, 2. 9 vols. Göttingen: Vandenhoeck & Ruprecht, 1963–1986.

Sigebert of Gembloux. *Chronica*. Ed. Ludwig Bethmann. MGH. Scriptores, 6. Hannover: Hahn, 1844.

Gaius Julius Solinus. *Collectanea rerum memorabilium*. Ed. Theodor Mommsen. 2nd ed. Berlin: Weidmann, 1958.

—. *The excellent and pleasant worke, Collectanea rerum memorabilium of Caius Julius Solinus, translated from the Latin (1587) by Arthur Golding. A facsimile reproduction with an introd. by George Kish*. Gainesville, FL: Scholars' Facsimiles & Reprints, 1955.

Stephanus Leodiensis. "L'office de l'Invention de S. Etienne." Ed. Ritva Jonsson. In *Historia. Études sur la genèse des offices versifiés*. Acta Universitatis Stockholmiensis. Studia Latina Stockholmiensia, 15. Stockholm: Almqvist & Wiksell, 1968.

Suetonius, *De uita Caesarum libri viii*. Ed. Maximilian Ihm. Editio minor. Stuttgart: Teubner, 1908; reprint 1967.

—. *Lives of the Caesars*. Trans. John C. Rolfe. 2 vols. LCL 31, 38. Cambridge, MA: Harvard University Press, 1914; reprint 1997–1998.

Suetonius/Donatus. *Vita Donati*. Ed. Colin Hardie. In *Vitae Vergilianae antiquae*. 2nd ed. Oxford: Clarendon, 1966. Pp. 1–18. Accessible at http://penelope.uchicago.edu/Thayer/L/Roman/Texts/Suetonius/de_Poetis/Vergil*.html (accessed August 5, 2011).

—. *Lives of Illustrious Men: Virgil*. In *Lives of the Caesars*. Trans. John C. Rolfe. Vol. 2. LCL 38. Cambridge, MA: Harvard University Press, 1914; reprint 1998.

Sulpicius Severus. *Epistolae*. Ed. Karl Halm. In *Libri qui supersunt*. CSEL 1. Vienna: Gerold, 1866. Pp. 138–51.

—. *Genuine Letters*. Trans. Alexander Roberts. NPNF2, 11.

—. *Vita s. Martini*. Ed. Karl Halm. In *Libri qui supersunt*. CSEL 1. Vienna: Gerold, 1866. Pp. 107–37.

—. *On the Life of St. Martin*. Trans. Alexander Roberts. NPNF2, 11.

Quintus Aurelius Symmachus. *Epistolae*. Ed. Otto Seeck. In *Opera quae supersunt*. MGH. Auctores antiquissimi, 6. Berlin: Weidmann, 1883. Pp. 1–317.

—. *Lettres*. Ed. and French trans. Jean Pierre Callu. 4 vols. Paris: Les Belles Lettres, 1972–2002.

Thomas Aquinas. *Catena aurea in quatuor Evangelia*. Ed. Angelico Guarienti. 2 vols. Turin: Marietti, 1953.

—. *Catena aurea: Commentary on the Four Gospels. Collected out of the Works of the Fathers by Saint Thomas Aquinas*. English trans. ed. John Henry Newman. 4 vols. Oxford: John Henry Parker; London: Rivington, 1841–1845; reprint Southampton, Engl.: Saint Austin Press, 1997.

—. *Scriptum super Sententiis*. In *Opera omnia*. 25 vols. Parma: Tipis Petri Fiaccadori, 1852–1873. Available at http://www.corpusthomisticum.org/snp4000.html (accessed December 10, 2010).

—. *Summa theologiae.* Blackfriars ed. 61 vols. London: Eyre and Spottiswoode; New York: McGraw-Hill, 1964–1973.

Thomas of Chobham. *Summa de commendatione virtutum et extirpatione vitiorum*. Ed. Franco Morenzoni. CCCM 82B. Turnhout: Brepols, 1997.

Thomas the Cistercian. *Commentarii in Cantica Canticorum*. PL 206:17A–862A.

Thomas de Eccleston. *Tractatus de adventu fratrum minorum in Angliam*. Ed. Andrew G. Little. Manchester: Manchester University Press, 1951.

—. *The Friars and How They Came to England*. Trans. Father Cuthbert. London: Sands, 1903.

Thomas of Ireland. *Manipulus florum*. Ed. Chris Nighman. Portions available at http://info.wlu.ca/~wwwhist/faculty/cnighman/index.html (accessed December 10, 2010).

Thomas de Radolio. *Vita Petri abbatis Claraevallis*. PL 209:1007A–36D.

Urso Salernitanus. *De commixtionibus elementorum libellus*. Ed. Wolfgang Stürner. Stuttgarter Beiträge zur Geschichte und Politik, 7. Stuttgart: Klett, 1976.

Valerius Maximus. *Factorum et dictorum memorabilium libri IX*. Ed. John Briscoe. 2 vols. Stuttgart and Lepzig: Teubner, 1998.

—. *Memorable Deeds and Sayings. One Thousand Tales from Ancient Rome*. Trans. Henry John Walker. Indianapolis, IN: Hackett, 2004.

Virgil. *Aeneid*. Ed. and trans. Henry Rushton Fairclough. 2nd rev. ed. Ed. George Patrick Goold. 2 vols. LCL 63–64. Cambridge, MA: Harvard University Press, 1935; reprint 1974.

Vincent of Beauvais. *De morali principis institutione*. Ed. Robert J. Schneider. CCCM 137. Turnhout: Brepols, 1995.

—. *De l'Institution morale du prince*. French trans. Charles Munier. Paris: Cerf, 2010.

—. *Speculum doctrinale*. In *Speculum quadruplex; sive Speculum maius*. 4 vols. Douai: Ex Officina Typographica B. Bellerie, 1624; reprint Graz: Akademische Druck- u. Verlaganstalt, 1964–1965. Vol. 2.

—. *Speculum historiale*. In *Speculum quadruplex; sive Speculum maius*. 4 vols. Douai:

Ex Officina Typographica B. Bellerie, 1624; reprint Graz: Akademische Druck- u. Verlaganstalt, 1964–1965. Vol. 4.

—. *Speculum naturale*. In *Speculum quadruplex; sive Speculum maius*. 4 vols. Douai: Ex Officina Typographica B. Bellerie, 1624; reprint Graz: Akademische Druck- u. Verlaganstalt, 1964–1965. Vol. 1.

Vita s. Audomari Episcopi Terwennensis. PL 147:1179C–1190B.

Vita beati Edmundi Cantuariensis Archiepiscopi et confessoris. In *Thesaurus novus anecdotorum*. Ed. Edmond Martène and Ursin Durand. 5 vols. Paris: F. Delaulne [etc.], 1717. Vol. 3. Cols. 1775–1826.

—. *The Life of St. Edmund*. By Matthew Paris. Trans. and ed. Clifford H. Lawrence. Oxford: A. Sutton, 1996.

Vita Secundi philosophi. Trans. Willelmus medicus. Ed. Walther Suchier. In *Altercatio Hadriani Augusti et Epicteti Philosophi*. Ed. Lloyd William Daly and Walther Suchier. Illinois Studies in Language and Literature, 24/1–2. Urbana, IL: University of Illinois Press, 1939. Reprint in Ben Edwin Perry. *Secundus the Silent Philosopher. The Greek Life of Secundus, Critically Edited and Restored so far as Possible, Together with Translations of the Greek and Oriental Versions, the Latin and Oriental Texts, and a Study of the Tradition*. Philological Monographs, 22. Ithaca, NY: The American Philological Association, 1964. Pp. 92–100.

Vitae patrum. PL 73:101A–74:240B.

—. *The Desert Fathers: Sayings of the Early Christian Monks*. Trans. Benedicta Ward. London and New York: Penguin, 2003. [*Vitae patrum*, Book 5]

—. *The Desert Fathers: Translations from the Latin*. Trans. Helen Waddell. London: Constable, 1936; reprint New York: Vintage Books, 1998.

Vitruvius. *De architectura libri decem*. Ed. Friedrich Krohn. Leipzig: Teubner, 1912.

—. *On Architecture*. Trans. Frank Granger. 2 vols. LCL 251, 280. Cambridge, MA: Harvard University Press, 1931–1934; reprint 1962.

—. *The Architecture of Marcus Vitruvius Pollio*. Trans. Joseph Gwilt. London: Priestley and Weale, 1826. Available at http://penelope.uchicago.edu/Thayer/E/Roman/Texts/Vitruvius/home.html (accessed August 5, 2011).

Walter of Châtillon. *Alexandreis*. Ed. Marvin L. Colker. Thesaurus mundi, 17. Padua: Antenore, 1978.

—. *The Alexandreis of Walter of Châtillon. A Twelfth-Century Epic*. Trans. David Townsend. Philadelphia: University of Pennsylvania Press, 1996.

William of Conches. *Dragmaticon Philosophiae*. Ed. Italo Ronca. CCCM 152. Turnhout: Brepols, 1997. Pp. 3–273.

—. *A Dialogue on Natural Philosophy (Dragmaticon philosophiae)*. Trans. Italo Ronca and Matthew Curr. Notre Dame Texts in Medieval Culture, 2. Notre Dame, IN: University of Notre Dame Press, 1996.

William of Saint-Thierry. *Epistola ad fratres de Monte Dei*. Ed. Volker Honemann. In *Die "Epistola ad fratres de Monte Dei" des Wilhelm von Saint-Thierry. Lateinische Überlieferung und mittelalterliche Übersetzungen*. Münchener Texte und Untersuchungen, 61. Munich and Zurich: Artemis, 1978. Pp. 288–413.

—. *The Golden Epistle: A Letter to the Brethren at Mont Dieu*. Trans. Theodore Berkeley. CFS 12. Kalamazoo, MI: Cistercian Publications, 1971; reprint 1980.

Secondary Material

Anger's Past: The Social Uses of an Emotion in the Middle Ages. Ed. Barbara Rosenwein. Ithaca and London: Cornell University Press, 1998.

Auerbach, Erich. "Figura." In *Scenes from the Drama of European Literature. Six Essays*. Trans. Ralph Manheim. New York: Meridian Books, 1959. Pp. 11–76.

Baker, Mona. *In Other Words: A Coursebook on Translation*. 2nd ed. London: Routledge, 2011.

Bataillon, Louis-Jacques. "Comptes de Pierre de Limoges pour la copie de livres." In *La production du livre universitaire au moyen âge: Exemplar et pecia, Actes du symposium tenu au Collegio San Bonaventura de Grottaferrata en mai, 1983*. Ed. Louis-Jacques Bataillon et al. Paris: Éditions du Centre national de la recherche scientifique, 1988. Pp. 265–73.

—. "Sermoni e orazioni d'ambiente universitario parigino nel sec. XIII." *Documenti e studi sulla tradizione filosofica medievale* 5 (1994): 297–329.

Baxandall, Michael. *Painting and Experience in Fifteenth Century Italy. A Primer in the Social History of Pictorial Style*. 2nd ed. Oxford and New York: Oxford University Press, 1988.

Bériou, Nicole. "Pierre de Limoges et la fin du temps." *Mélanges de l'École française de Rome* (Série Moyen Age et Temps Modernes) 98,1 (1986): 65–107.

—. "La prédication au béguinage de Paris pendant l'année liturgique 1272–1273." *Recherches augustiniennes* 13 (1978): 105–229.

Berlioz, Jacques. "Etienne de Bourbon, *Tractatus de diversis materiis predicabilibus*." In Jacques Berlioz and Marie Anne Polo de Beaulieu. *Les Exempla médiévaux*. Carcassonne: n.p., n.d. Pp. 135–49.

—. "Virgile dans la littérature des *exempla* (XIIIe–XVe siècles)." In *Lectures médiévales de Virgile. Actes du Colloque organisé par l'École française de Rome (Rome, 25–28 octobre 1982)*. Collection de l'École française de Rome, 80. Paris and Turin: École française de Rome, 1985. Pp. 65–120.

Biernoff, Suzannah. *Sight and Embodiment in the Middle Ages*. Houndmills and New York: Palgrave Macmillan, 2002.

Bloomfield, Morton W. *The Seven Deadly Sins: An Introduction to the History of a Religious Concept, with Special Reference to Medieval English Literature*. [East Lansing]: Michigan State University Press, 1952; reprint 1967.

Casagrande, Carla, and Silvana Vecchio. *I peccati della lingua. Disciplina ed etica della parola nella cultura medievale*. Bibliotheca Biographica. Sezione Storico-Antropologica. Rome: Istituto della Enciclopedia Italiana, 1987.

—. *I sette vizi capitali. Storia dei peccati nel Medioevo*. Saggi, 832. Turin: Einaudi, 2000.

Cavadini, John C. "Pride." In *Augustine through the Ages: An Encyclopedia*. Ed. Allan Fitzgerald. Grand Rapids, MI: William B. Eerdmans, 1999. Pp. 679–84.

Clark, David L. "Optics for Preachers: The *De oculo morali* by Peter of Limoges." *Michigan Academician* 9 (1977): 329–43.

Clark, Willene B. "Commentary." In *A Medieval Book of Beasts. The Second-Family Bestiary*. Ed. and trans. Willene B. Clark. Woodbridge: Boydell Press, 2006. Pp. 7–116.

Craun, Edwin D. *Lies, Slander, and Obscenity in Medieval English Literature: Pastoral Rhetoric and the Deviant Speaker*. Cambridge Studies in Medieval Literature, 31. Cambridge, Engl.: Cambridge University Press, 1997.

d'Alverny, Marie-Thérèse. "Un adversaire de Saint Thomas: Petrus Iohannis Olivi." In *St. Thomas Aquinas 1274–1974: Commemorative Studies*. Ed. Armand A. Maurer et al. 2 vols. Toronto: Pontifical Institute of Mediaeval Studies, 1974. Vol. 2. Pp. 179–218.

Davis, Henry W.C. "An Unpublished Life of Edmund Rich." *The English Historical Review* 22, no. 85 (1907): 84–92.

Delisle, Léopold. *Le Cabinet des Manuscrits de la Bibliothèque impériale*. 3 vols. Histoire Générale de Paris, 6. Paris: Imprimerie impériale, 1868–1881; reprint New York: Burt Franklin, 1973.

Denery, Dallas G., II. *Seeing and Being Seen in the Later Medieval World. Optics, Theology and Religious Life*. Cambridge Studies in Medieval Life and Thought, 4th ser., 63. Cambridge, Engl.: Cambridge University Press, 2005.

Diekstra, F.N.M. "The Art of Denunciation: Medieval Moralists on Envy and Detraction." In *In the Garden of Evil*, ed. Newhauser. Pp. 431–54.

du Cange, Charles du Fresne, Sieur. *Glossarium mediae et infimae latinitatis*. Ed. nova. Niort: L. Favre, 1883–1887; reprint 5 vols., Graz: Akademische Druck- u. Verlagsanstalt, 1954.

Emery, Kent, Jr., and Andreas Speer. "After the Condemnation of 1277: New Evidence, New Perspectives, and Grounds for New Interpretations." In *Nach der Verurteilung von 1277. Philosophie und Theologie an der Universität von Paris im letzten Viertel des 13. Jahrhunderts. Studien und Texte / After the Condemnation of 1277. Philosophy and Theology at the University of Paris in the Last Quarter of the Thirteenth Century. Studies and Texts*. Ed. Jan A. Aertsen, Kent Emery, Jr., and Andreas Speer. Miscellanea Mediaevalia, 28. Berlin and New York: Walter de Gruyter, 2001. Pp. 3–19.

Evans, Gillian R. *Getting it Wrong. The Medieval Epistemology of Error*. Studien und Texte zur Geistesgeschichte des Mittelalters, 63. Leiden, etc.: Brill, 1998.

Ferruolo, Stephen C. *The Origins of the University: The Schools of Paris and Their Critics, 1100–1215*. Stanford: Stanford University Press, 1985.

Forcellini, Egidio. *Lexicon totius latinitatis*. Rev. ed. Ed. Giuseppe Furlanetto, Francesco Corradini, and Giuseppe Perin. 6 vols. Padua: [Typis Seminarii], 1864–1898; reprint 1940.

Friedenthal, Meelis. *Tallinna Linnaarhiivi: Tractatus moralis de oculo*. Dissertationes Theologiae Universitatis Tartuensis, 13. Tartu: Ülikooli Kirjastus, 2008.

Glorieux, Palémon. *Aux origines de la Sorbonne, I: Robert de Sorbon: l'homme, le collège, les documents*. Études de philosophie médiévale, 53. Paris: Vrin, 1966.

—. *Répertoire des maîtres en théologie de Paris au XIIIe siècle*. 2 vols. Études de philosophie médiévale, 17–18. Paris: Vrin, 1933–1934.

Godman, Peter. *The Silent Masters: Latin Literature and Its Censors in the High Middle Ages*. Princeton: Princeton University Press, 2000.

Hain, Ludwig. *Repertorium bibliographicum*. 2 vols. Stuttgart: J. G. Cotta, 1826–1838; reprint Berlin: J. Altmann, 1925.

Hauréau, Bernard. "Sermonnaires." In *Histoire littéraire de la France*. Vol. 26. Paris: Imprimerie nationale, 1873. Pp. 387–468.

Hempel, Wolfgang. *Übermuot diu alte .... Der Superbia-Gedanke und seine Rolle in der deutschen Literatur des Mittelalters*. Studien zur Germanistik, Anglistik und Komparatistik, 1. Bonn: H. Bouvier, 1970.

In the Garden of Evil: The Vices and Culture in the Middle Ages. Ed. Richard Newhauser. Papers in Mediaeval Studies, 18. Toronto: Pontifical Institute of Mediaeval Studies, 2005.

Janson, Horst W. *Apes and Ape Lore in the Middle Ages and the Renaissance*. Studies of the Warburg Institute, 20. London: The Warburg Institute, 1952.

Karras, Ruth Mazo. "The Latin Vocabulary of Illicit Sex in English Ecclesiastical Court Records." *Journal of Medieval Latin* 2 (1992): 1–17.

—. "The Lechery That Dare Not Speak Its Name: Sodomy and the Vices in Medieval England." In *In the Garden of Evil*, ed. Newhauser, 193–205.

Kenny, Dorothy. "Equivalence." In *Routledge Encyclopedia of Translation Studies*. 2nd ed. Ed. Mona Baker and Gabriela Saldanha. London and New York: Routledge, 2009. Pp. 96–99.

Kenny, Neil. *The Uses of Curiosity in Early Modern France and Germany*. Oxford: Oxford University Press, 2004.

Kessler, Herbert L. "*Speculum*." *Speculum* 86,1 (2011): 1–41.

Kienzle, Beverly Mayne. "Cistercian Views of the City in the Sermons of Hélinand of Froidmont." In *Medieval Sermons and Society: Cloister, City, University. Proceedings of the International Symposia at Kalamazoo and New York*. Ed. Jaqueline Hamesse et al. Textes et études du moyen âge, 9. Louvain-la-Neuve: F.I.D.E.M., 1998. Pp. 165–82.

Kitchen, John. "Cassian, Nocturnal Emissions, and the Sexuality of Jesus." In *The Seven Deadly Sins*. Ed. Newhauser, 73–94.

Lauer, Hans H. "Qusta ibn Luqa." In *Lexikon des Mittelalters*. 10 vols. Munich and Zurich: Artemis, 1980–1999. Vol. 7. Col. 378.

Lausberg, Heinrich. "Ave maris stella." In *Lexikon für Theologie und Kirche*. 2nd ed. 10 vols. Freiburg i. Br.: Herder, 1957–1965. Vol. 1. Cols. 1141–42.

Lecoy de la Marche, Albert. *La chaire française au moyen âge*. Paris: Didier, 1886.

Lector et compilator. Vincent de Beauvais, frère prêcheur: un intellectuel et son milieu au XIIIe siècle. Ed. Serge Lusignan and Monique Paulmier-Foucart, with Marie-Christine Duchenne. Grâne, France: Créaphis, 1997.

Lindberg, David C. "Alhazen's Theory of Vision and Its Reception in the West."

Isis 58 (1967): 321–41. Reprint in David C. Lindberg. *Studies in the History of Medieval Optics*. III.

—. *The Beginnings of Western Science. The European Scientific Tradition in Philosophical, Religious, and Institutional Context, 600 B.C. to A.D. 1450*. 2nd ed. Chicago and London: University of Chicago Press, 2007.

—. "The Science of Optics." In *Science in the Middle Ages*. Ed. David C. Lindberg. Chicago and London: University of Chicago Press, 1978. Pp. 338–68. Reprint in David C. Lindberg. *Studies in the History of Medieval Optics*. I.

—. *Studies in the History of Medieval Optics*. Variorum Reprints. Collected Studies Series CS186. London: Ashgate, 1983.

—. *Theories of Vision from Al-Kindi to Kepler*. Chicago and London: University of Chicago Press, 1976.

Little, Lester K. "Pride Goes before Avarice: Social Change and the Vices in Latin Christendom." *The American Historical Review* 76 (1971): 16–49.

Longère, Jean. *La prédication médiévale*. Paris: Études augustiniennes, 1983.

Luna, Concetta. *Aegidii Romani opera omnia, 1.1/3**: Catalogo dei manoscritti (294–372): Francia (Parigi)*. Testi e studi per il "Corpus Philosophorum Medii Aevi," 7. Florence: Leo S. Olschki, 1988.

Mabille, Madeleine. "Pierre de Limoges, copiste de manuscrits." *Scriptorium* 24 (1970): 45–47 and illus. 10–13.

—. "Pierre de Limoges et ses méthodes de travail." In *Hommages à André Boutemy*. Ed. Guy Cambier. Collection Latomus, 145. Brussels: Revue d'études Latines, 1976. Pp. 244–51.

Mariaux, Pierre Alain. "Collecting (and Display)." In *A Companion to Medieval Art: Romanesque and Gothic in Northern Europe*. Ed. Conrad Rudolph. Blackwell Companions to Art History, 2. Malden, MA: Blackwell, 2006. Pp. 213–32.

Martin, Charles Trice. "Preface." In *Registrum epistolarum fratris Johannis Peckham, archiepiscopi cantuariensis. ... Published by the authority of the lords commissioners of Her Majesty's Treasury, under the direction of the master of the rolls*. Ed. Charles Trice Martin. Rerum Britannicarum Medii Aevi scriptores, or Chronicles and memorials of Great Britain and Ireland during the Middle Ages, 77. 3 vols. London: Longman & Co., 1882–1885; reprint [Nendeln]: Kraus, 1965. Vol. 3. Pp. xxvii–cxlv.

Murano, Giovanna. *Opere diffuse per exemplar e pecia*. Textes et études du moyen âge, 29. Turnhout: Brepols, 2005.

Murray, Alexander. *Reason and Society in the Middle Ages*. Oxford: Clarendon, 1978.

Musurillo, Herbert. "The Mediaeval Hymn, *Alma Redemptoris*: A Linguistic Analysis." *The Classical Journal* 52,4 (1957): 171–74.

Newhauser, Richard. *The Early History of Greed: The Sin of Avarice in Early Medieval Thought and Literature*. Cambridge Studies in Medieval Literature, 41. Cambridge, Engl.: Cambridge University Press, 2000; reprint 2006.

—. "*Inter scientiam et populum*: Roger Bacon, Peter of Limoges, and the 'Tractatus moralis de oculo.'" In *Nach der Verurteilung von 1277. Philosophie und Theologie an der Universität von Paris im letzten Viertel des 13. Jahrhunderts. Studien und Texte / After the Condemnation of 1277. Philosophy and Theology at the University of Paris in the Last Quarter of the Thirteenth Century. Studies and Texts*. Ed. Jan A. Aertsen, Kent Emery, Jr., and Andreas Speer. Miscellanea Mediaevalia, 28. Berlin and New York: Walter de Gruyter, 2001. Pp. 682–703.

—. "A Middle English Poem on the Fleeting Nature of Material Wealth." *Medium Ævum* 71,1 (2002): 74–81.

—. "Nature's Moral Eye: Peter of Limoges' *Tractatus moralis de oculo*." In *Man and Nature in the Middle Ages*. Ed. Susan J. Ridyard and Robert G. Benson. Sewanee Mediaeval Studies, 6. Sewanee, TN: The University of the South Press, 1995. Pp. 125–36.

—. "Peter of Limoges, Optics, and the Science of the Senses." In *Pleasure and Danger in Perception: The Five Senses in the Middle Ages and the Renaissance*; special issue of *The Senses & Society* 5,1 (2010): 28–44.

—. "Towards a History of Human Curiosity: A Prolegomenon to Its Medieval Phase," *Deutsche Vierteljahrsschrift* 56 (1982): 559–75. Reprint in Richard Newhauser. *Sin: Essays on the Moral Tradition in the Western Middle Ages*. Variorum Collected Studies Series, CS869. Aldershot, Engl., and Burlington, VT: Ashgate Publishing, 2007. Essay XIII.

—. "Der 'Tractatus moralis de oculo' des Petrus von Limoges und seine *exempla*." In *Exempel und Exempelsammlungen*. Fortuna vitrea, 2. Ed. Walter Haug and Burghart Wachinger. Tübingen: Niemeyer, 1991. Pp. 95–136.

—. *The Treatise on Vices and Virtues in Latin and the Vernacular*. Typologie des sources du moyen âge occidental, 68. Turnhout: Brepols, 1993.

—, and István Bejczy. *A Supplement to Morton W. Bloomfield et al., 'Incipits of Latin Works on the Virtues and Vices, 1100–1500 A.D.'* Instrumenta Patristica et Mediaevalia, Research on the Inheritance of Early and Medieval Christianity, 50. Turnhout: Brepols, 2008.

—, Tiina Kala, and Meelis Friedenthal. "The Work of an English Scribe in a Manuscript in Estonia." *Scriptorium* 62,1 (2008): 139–48, plate 19.

Nichols, Stephen G. "'The Pupil of Your Eye': Vision, Language, and Poetry in Thirteenth-Century Paris." In *Rethinking the Medieval Senses. Heritage, Fascinations, Frames*. Ed. Stephen G. Nichols, Andreas Kablitz, and Alison Calhoun. Baltimore: The Johns Hopkins University Press, 2008. Pp. 286–307.

Paulmier-Foucart, Monique. With Marie-Christine Duchenne. *Vincent de Beauvais et le grand miroir du monde.* Témoins de notre histoire, 10. Turnhout: Brepols, 2004.

Payer, Pierre J. *The Bridling of Desire: Views of Sex in the Later Middle Ages*. Toronto, etc.: University of Toronto Press, 1993.

—. *Sex and the New Medieval Literature of Confession, 1150–1300*. Studies and Texts, 163. Toronto: Pontifical Institute of Mediaeval Studies, 2009.

Peters, Edward. "*Libertas Inquirendi* and the *Vitium Curiositatis* in Medieval Thought." In *La Notion de liberté au Moyen Âge. Islam, Byzance, Occident: Penn-*

Paris-Dumbarton Oaks Colloquia, IV. Paris: Société d'Édition Les Belles Lettres, 1985. Pp. 89–98.

Roger Bacon and the Sciences: Commemorative Essays. Ed. Jeremiah Hackett. Studien und Texte zur Geistesgeschichte des Mittelalters, 57. Leiden and New York: Brill, 1997.

Rouse, Richard H. "The Early Library of the Sorbonne." *Scriptorium* 21 (1967): 42–71, 227–51. Reprint in *Authentic Witnesses*. Pp. 341–408.

Rouse, Richard H., and Mary A. Rouse. *Authentic Witnesses. Approaches to Medieval Texts and Manuscripts*. Publications in Medieval Studies, 17 [27]. Notre Dame: University of Notre Dame Press, 1991.

—. "The Book Trade at the University of Paris, ca. 1250–ca. 1350." In *La production du livre universitaire au moyen âge: Exemplar et pecia, Actes du symposium tenu au Collegio San Bonaventura de Grottaferrata en mai, 1983*. Ed. Louis-Jacques Bataillon et al. Paris: Éditions du Centre national de la recherche scientifique, 1988. Pp. 41–114. Reprint in *Authentic Witnesses*. Pp. 259–338.

—. *Manuscripts and Their Makers. Commercial Book Producers in Medieval Paris 1200–1500*. 2 vols. Turnhout: Harvey Miller, 2000.

—. *Preachers, Florilegia, and Sermons: Studies on the Manipulus florum of Thomas of Ireland*. Studies and Texts, 47. Toronto: Pontifical Institute of Mediaeval Studies, 1979.

Sabra, Abdelhamid I. "Psychology versus Mathematics: Ptolemy and Alhazen on the Moon Illusion." In *Mathematics and Its Applications to Science and Natural Philosophy in the Middle Ages. Essays in Honor of Marshall Clagett*. Ed. Edward Grant and John E. Murdoch. Cambridge, Engl.: Cambridge University Press, 1987. Pp. 217–47.

Sarton, George. *Introduction to the History of Science*. 3 vols. Publications of the Carnegie Institution of Washington, 376. Baltimore: Published for the Carnegie Institution of Washington, by the Williams & Wilkins Company, 1927–1947.

Schleusener-Eichholz, Gudrun. *Das Auge im Mittelalter*. Münstersche Mittelalter-Schriften, 35. Munich: Fink, 1985.

—. "Naturwissenschaft und Allegorese: Der 'Tractatus de oculo morali' des Petrus von Limoges." *Frühmittelalterliche Studien* 12 (1978): 258–309.

The Seven Deadly Sins: From Communities to Individuals. Ed. Richard Newhauser. Studies in Medieval and Reformation Traditions: History, Culture, Religion, Ideas, 123. Leiden and Boston: Brill, 2007.

Shooner, Hugues V. "La production du livre par la pecia." In *La production du livre universitaire au moyen âge: Exemplar et pecia, Actes du symposium tenu au Collegio San Bonaventura de Grottaferrata en mai, 1983*. Ed. Louis-Jacques Bataillon et al. Paris: Éditions du Centre national de la recherche scientifique, 1988. Pp. 17–37.

Smith, A. Mark. "Getting the Big Picture in Perspectivist Optics." *Isis* 72 (1981): 568–89.

Soetermeer, Frank P.W. *Utrumque ius in peciis. Aspetti della produzione libraria a*

Bologna fra due e trecento. Orbis academicus: Saggi e documenti di storia delle università, 7. Milan: Giuffrè, 1997.

Soler, Albert. "Ramon Llull and Peter of Limoges." *Traditio* 48 (1993): 93–105.

Spettmann, Hieronymus. "Das Schriftchen 'De oculo morali' und sein Verfasser." *Archivum Franciscanum Historicum* 16 (1923): 309–22.

Straw, Carole. "Gregory, Cassian, and the Cardinal Vices." In *In the Garden of Evil*, ed. Newhauser, 35–58.

Tachau, Katherine H. "Seeing as Action and Passion in the Thirteenth and Fourteenth Centuries." In *The Mind's Eye. Art and Theological Argument in the Middle Ages*. Ed. Jeffrey F. Hamburger and Anne-Marie Bouché. Princeton: Department of Art and Archaeology, Princeton University, and Princeton University Press, 2006. Pp. 336–59.

Teetaert, Amédée. "Jean Pecham." In *Dictionnaire de théologie catholique*. 15 vols. Paris: Letouzey et Ané, 1908–1950. Vol. 12/1. Cols. 100–40.

Theunissen, Michael. *Vorentwürfe der Moderne: Antike Melancholie und die Acedia des Mittelalters*. Berlin and New York: Walter de Gruyter, 1996.

Thorndike, Lynn. "Peter of Limoges on the Comet of 1299." *Isis* 36 (1945): 3–6.

Tobler, Alfred. *Tobler-Lommatzsch, altfranzösisches Wörterbuch. Adolf Toblers nachgelassene Materialien*. Ed. Erhard Lommatzsch. 11 vols. Berlin: Weidmann, 1925–2002.

Vincent-Cassy, Mireille. "Between Sin and Pleasure: Drunkenness in France in the Late Middle Ages." Trans. Erika Pavelka. In *In the Garden of Evil*, ed. Newhauser, 393–430.

—. "L'envie au moyen âge." *Annales. Économies, Sociétés, Civilisations* 35 (1980): 253–71.

—. "La *gula* curiale ou les débordements des banquets au début du règne de Charles VI." In *La Sociabilité à table: Commensalité et convivialité à travers les âges*. Ed. Martin Aurell, Olivier Dumoulin, and Françoise Thelamon. Publications de l'Université de Rouen, 178. [Rouen]: Publications de l'Université de Rouen, 1992. Pp. 91–102.

Weijers, Olga. *Terminologie des universités au XIIIe siècle*. Lessico intellettuale europeo, 39. Rome: Edizioni dell'Ateneo, 1987.

Wenzel, Siegfried. "The Continuing Life of William Peraldus's *Summa vitiorum*." In *Ad litteram. Authoritative Texts and Their Medieval Readers*. Ed. Mark D. Jordan and Kent Emery, Jr. Notre Dame and London: University of Notre Dame Press, 1992. Pp. 135–63.

—. *Latin Sermon Collections from Later Medieval England: Orthodox Preaching in the Age of Wyclif*. Cambridge Studies in Medieval Literature, 53. Cambridge, Engl.: Cambridge University Press, 2005.

—. *The Sin of Sloth: Acedia in Medieval Thought and Literature*. Chapel Hill: University of North Carolina Press, 1967.

Williams, Steven J. *The Secret of Secrets. The Scholarly Career of a Pseudo-Aristotelian Text in the Latin Middle Ages*. Ann Arbor: University of Michigan Press, 2003.

Index of Sources and Analogues

Abbreviated titles of works used in the footnotes are given here in parentheses after the full titles. In sections B and C, anonymous works are listed first.

A. The Vulgate Bible

B. Christian and Muslim Authors and Texts

C. Classical, Pseudo-Classical, and Late-Antique Authors and Texts

Index of Names

The following index contains names occurring in the introduction and translation and mentioned incidentally in the notes. Authors and titles of works cited in the footnotes to the translation are found in the Index of Sources and Analogues.

Index of Subjects